Holt Mathematics

Chapter 2 Resource Book

HOLT, RINEHART AND WINSTON

A Harcourt Education Company

Orlando • Austin • New York • San Diego • London

Copyright © by Holt, Rinehart and Winston

All rights reserved. No part of this publication may be reproduced or transmitted in any form or by any means, electronic or mechanical, including photocopy, recording, or any information storage and retrieval system, without permission in writing from the publisher.

Teachers using HOLT MATHEMATICS may photocopy complete pages in sufficient quantities for classroom use only and not for resale.

Printed in the United States of America

If you have received these materials as examination copies free of charge, Holt, Rinehart and Winston retains title to the materials and they may not be resold. Resale of examination copies is strictly prohibited and is illegal.

Possession of this publication in print format does not entitle users to convert this publication, or any portion of it, into electronic format.

ISBN 0-03-078391-7

6 7 8 9 10   170   10 09 08

CONTENTS

Blackline Masters

Parent Letter	1
Lesson 2-1 Practice A, B, C	3
Lesson 2-1 Reteach	6
Lesson 2-1 Challenge	8
Lesson 2-1 Problem Solving	9
Lesson 2-1 Reading Stratigies	10
Lesson 2-1 Puzzles, Twisters & Teasers	11
Lesson 2-2 Practice A, B, C	12
Lesson 2-2 Reteach	15
Lesson 2-2 Challenge	16
Lesson 2-2 Problem Solving	17
Lesson 2-2 Reading Strategies	18
Lesson 2-2 Puzzles, Twisters & Teasers	19
Lesson 2-3 Practice A, B, C	20
Lesson 2-3 Reteach	23
Lesson 2-3 Challenge	24
Lesson 2-3 Problem Solving	25
Lesson 2-3 Reading Strategies	26
Lesson 2-3 Puzzles, Twisters & Teasers	27
Lesson 2-4 Practice A, B, C	28
Lesson 2-4 Reteach	31
Lesson 2-4 Challenge	32
Lesson 2-4 Problem Solving	33
Lesson 2-4 Reading Strategies	34
Lesson 2-4 Puzzles, Twisters, & Teasers	35
Lesson 2-5 Practice A, B, C	36
Lesson 2-5 Reteach	39
Lesson 2-5 Challenge	40
Lesson 2-5 Problem Solving	41
Lesson 2-5 Reading Strategies	42
Lesson 2-6 Exploration Recording Sheet	43
Lesson 2-6 Practice A, B, C	44
Lesson 2-6 Reteach	47
Lesson 2-6 Challenge	49
Lesson 2-6 Problem Solving	50
Lesson 2-6 Reading Strategies	51
Lesson 2-6 Puzzles, Twisters & Teasers	52
Lesson 2-7 Practice A, B, C	53
Lesson 2-7 Reteach	56
Lesson 2-7 Challenge	57
Lesson 2-7 Problem Solving	58
Lesson 2-7 Reading Strategies	59
Lesson 2-7 Puzzles, Twisters & Teasers	60
Lesson 2-8 Practice A, B, C	61
Lesson 2-8 Reteach	64
Lesson 2-8 Challenge	66
Lesson 2-8 Problem Solving	67
Lesson 2-8 Reading Strategies	68
Lesson 2-8 Puzzles, Twisters & Teasers	69
Teacher Tool: Cutouts	70
Answers to Blackline Masters	71

Date _____

Dear Family,

In this chapter, your child will learn the definition of rational numbers, how to perform operations involving rational numbers, and how to solve equations using them.

Rational numbers are the set of numbers that can be written in the form $\frac{n}{d}$, where n and d are integers and d does not equal zero. Decimals that terminate or repeat are also rational numbers.

Numerator ⟶ $\frac{n}{d}$ ⟵ Denominator

Fractions can be simplified by dividing the **numerator** and **denominator** by the same nonzero integer.

For example, this is how you **simplify** $\frac{12}{15}$.
$$\frac{12}{15} = \frac{12 \div 3}{15 \div 3} = \frac{4}{5}$$

Your child will learn to add, subtract, multiply, and divide rational numbers with **like** and **unlike denominators**.

This is how you add fractions with *like* denominators, such as $\frac{6}{11} + \frac{9}{11}$.

$\frac{6}{11} + \frac{9}{11}$ Add the numerators: $6 + 9 = 15$.

$\frac{15}{11}$ Keep the denominator.

This is how you add fractions with *unlike* denominators, such as $\frac{2}{3} + \frac{1}{5}$.

$\frac{2}{3} + \frac{1}{5}$ *Find the common denominator: 3(5) = 15.*

$= \frac{2}{3} \cdot \frac{5}{5} + \frac{1}{5} \cdot \frac{3}{3}$ *Multiply by fractions equal to 1.*

$= \frac{10}{15} + \frac{3}{15}$ *Rewrite using common denominators.*

$= \frac{13}{15}$ *Add.*

Your child will solve one-step equations with decimals and fractions, such as $y - 12.5 = 17$.

$$\begin{array}{r} y - 12.5 = 17 \\ +12.5 \quad +12.5 \\ \hline y = 29.5 \end{array}$$ Add 12.5 to both sides.

Holt Mathematics

Your child will also solve two-step equations with rational numbers. As the name implies, two-step equations require two-steps to solve.

One type of two-step equation is $1.5x + 4.3 = 6.25$.
To solve this equation, isolate the variable by subtracting first and then dividing.

$$1.5x + 4.3 = 6.25$$
$$\; -4.3 \quad -4.30 \quad \text{Subtract 4.3 from both sides.}$$
$$1.5x = 1.95 \quad \text{Simplify.}$$
$$\frac{1.5x}{1.5} = \frac{1.95}{1.5} \quad \text{Divide both sides by 1.5}$$
$$x = 1.3 \quad \text{Simplify.}$$

Another type of two-step equation is $\frac{y-4}{3} = -1$.

To solve this equation, isolate the variable by multiplying first and then adding.

$$\frac{y-4}{3} = -1$$
$$(3)\frac{y-4}{3} = -1(3) \quad \text{Multiply both sides by 3.}$$
$$y - 4 = -3 \quad \text{Simplify.}$$
$$\; +4 \quad +4 \quad \text{Add 4 to both sides.}$$
$$y = 1 \quad \text{Simplify.}$$

Your child will use both one-step and two-step equations with rational numbers to solve real-life problems.

For additional resources, visit go.hrw.com and enter the keyword MT7 Parent.

Name _____ Date _____ Class _____

LESSON 2-1 Practice A
Rational Numbers

Simplify.

1. $\dfrac{4}{12}$ 2. $\dfrac{5}{15}$ 3. $-\dfrac{2}{8}$ 4. $\dfrac{6}{24}$

_____ _____ _____ _____

5. $\dfrac{14}{24}$ 6. $-\dfrac{15}{35}$ 7. $\dfrac{10}{21}$ 8. $-\dfrac{16}{36}$

_____ _____ _____ _____

Write each decimal as a fraction in simplest form.

9. 0.4 10. −0.35 11. 0.105 12. 1.2

_____ _____ _____ _____

13. −0.85 14. 0.325 15. 0.002 16. 2.3

_____ _____ _____ _____

17. 0.28 18. −1.25 19. 0.064 20. 0.0075

_____ _____ _____ _____

Write each fraction as a decimal.

21. $\dfrac{1}{9}$ 22. $\dfrac{9}{16}$ 23. $-\dfrac{11}{20}$ 24. $\dfrac{6}{5}$

_____ _____ _____ _____

25. $\dfrac{2}{15}$ 26. $-2\dfrac{7}{12}$ 27. $\dfrac{3}{100}$ 28. $5\dfrac{8}{25}$

_____ _____ _____ _____

29. Make up a fraction that cannot be simplified that has 12 as its denominator.

_____

Copyright © by Holt, Rinehart and Winston.
All rights reserved.

Holt Mathematics

Name _____ Date _____ Class _____

LESSON 2-1 Practice B
Rational Numbers

Simplify.

1. $\dfrac{6}{9}$
2. $\dfrac{48}{96}$
3. $\dfrac{13}{52}$
4. $-\dfrac{7}{28}$

5. $\dfrac{15}{40}$
6. $-\dfrac{4}{48}$
7. $-\dfrac{14}{63}$
8. $\dfrac{12}{72}$

Write each decimal as a fraction in simplest form.

9. 0.72
10. 0.058
11. −1.65
12. 2.1

13. 0.036
14. −4.06
15. 2.305
16. 0.0064

17. −0.60
18. 6.95
19. 0.016
20. 0.0005

Write each fraction as a decimal.

21. $\dfrac{1}{8}$
22. $\dfrac{8}{3}$
23. $\dfrac{14}{15}$
24. $\dfrac{16}{5}$

25. $\dfrac{11}{16}$
26. $\dfrac{7}{9}$
27. $\dfrac{4}{5}$
28. $\dfrac{31}{25}$

29. Make up a fraction that cannot be simplified that has 24 as its denominator.

Name _____ Date _____ Class _____

LESSON 2-1 Practice C
Rational Numbers

Write each decimal as a fraction in simplest form.

1. 0.9
2. 2.5
3. −0.36
4. −0.215

_____   _____   _____   _____

5. −4.02
6. 0.0085
7. 1.006
8. 0.45

_____   _____   _____   _____

Write each fraction as a decimal.

9. $\frac{9}{15}$
10. $-\frac{22}{50}$
11. $\frac{45}{16}$
12. $-\frac{18}{90}$

_____   _____   _____   _____

13. $\frac{15}{80}$
14. $\frac{21}{126}$
15. $\frac{19}{12}$
16. $\frac{39}{20}$

_____   _____   _____   _____

17. Make up a fraction that cannot be simplified that has 48 as its denominator.

_____

18. a. Simplify each fraction below.

   b. Write the denominator of each simplified fraction as the product of prime factors.

   c. Write each fraction as a decimal. Label each as a terminating or repeating decimal.

   $\frac{4}{36}$ $\frac{5}{40}$ $\frac{10}{25}$

   a. _____   a. _____   a. _____

   b. _____   b. _____   b. _____

   c. _____   c. _____   c. _____

Copyright © by Holt, Rinehart and Winston.
All rights reserved.

Holt Mathematics

Name _____ Date _____ Class _____

LESSON 2-1 Reteach
Rational Numbers

A **rational number** is a *ratio* of two integers.

Rational Number = $\dfrac{\text{Integer}}{\text{Integer}}$ ← Numerator
← Denominator

The set of rational numbers contains:
all integers
all fractions
decimals that repeat, such as $0.4\overline{6}$
decimals that terminate, such as 3.5

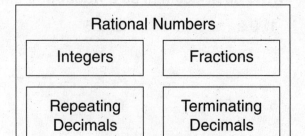

To simplify a fraction, divide numerator and denominator by the highest common factor.

$\dfrac{5}{15} = \dfrac{5 \div 5}{15 \div 5} = \dfrac{1}{3}$

Complete to simplify each fraction.

1. $\dfrac{8}{16} = \dfrac{8 \div 8}{16 \div 8} = $ _____

2. $\dfrac{15}{45} = \dfrac{\div}{\div} = $ _____

3. $\dfrac{12}{30} = \dfrac{\div}{\div} = $ _____

4. $\dfrac{12}{24} = \dfrac{\div}{\div} = $ _____

5. $\dfrac{5}{35} = \dfrac{\div}{\div} = $ _____

6. $\dfrac{14}{49} = \dfrac{\div}{\div} = $ _____

Simply each fraction.

7. $\dfrac{8}{56} = $ _____

8. $\dfrac{15}{50} = $ _____

9. $\dfrac{8}{36} = $ _____

To write a decimal as a fraction, use the number of decimal places to get the denominator. Then simplify.

$0.4 = \dfrac{4}{10} = \dfrac{4 \div 2}{10 \div 2} = \dfrac{2}{5}$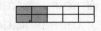

Complete to write each decimal as a fraction in simplest form.

10. $0.25 = \dfrac{25}{100} = \dfrac{25 \div}{100 \div} = $ _____

11. $0.375 = \dfrac{375}{1000} = \dfrac{375 \div}{1000 \div} = $ _____

Write each decimal as a fraction in simplest form.

12. $0.55 = $ _____

13. $0.32 = $ _____

Reteach
2-1 Rational Numbers (continued)

To write a fraction as a decimal, divide numerator by denominator.

A decimal may terminate.

$$\frac{3}{4} = 4\overline{)3.00}$$
$$0.75$$
$$-28\downarrow$$
$$20$$
$$-20$$
$$0$$

A decimal may repeat.

$$\frac{1}{3} = 3\overline{)1.00}$$
$$0.\overline{3}$$
$$-9\downarrow$$
$$10$$
$$-9$$
$$1$$

Complete to write each fraction as a decimal.

14. $\frac{15}{4} = 4\overline{)15.00}$

15. $\frac{5}{6} = 6\overline{)5.00}$

16. $\frac{11}{3} = 3\overline{)11.00}$

Write each fraction as a decimal.

17. $\frac{5}{2} =$ _____

18. $\frac{15}{8} =$ _____

19. $\frac{28}{6} =$ _____

20. $\frac{22}{4} =$ _____

21. $\frac{62}{12} =$ _____

22. $\frac{105}{10} =$ _____

Name _____ Date _____ Class _____

LESSON 2-1 Challenge
Encore, Encore, ...

Explore some patterns with repeating decimals. Use a calculator to write each decimal equivalent.

1. $\dfrac{1}{9} = $ _____ 2. $\dfrac{2}{9} = $ _____ 3. $\dfrac{3}{9} = $ _____

Predict the decimal equivalent of each fraction. Verify your results on a calculator.

4. $\dfrac{4}{9} = $ _____ 5. $\dfrac{6}{9} = $ _____ 6. $\dfrac{8}{9} = $ _____

Write each fractional equivalent.

7. $0.\overline{5} = $ _____ 8. $0.\overline{7} = $ _____ 9. $0.\overline{9} = $ _____

Use a calculator to write each decimal equivalent.

10. $\dfrac{42}{99} = $ _____ 11. $\dfrac{358}{999} = $ _____ 12. $\dfrac{4276}{9999} = $ _____

Predict the decimal equivalent of each fraction.

13. $\dfrac{76}{99} = $ _____ 14. $\dfrac{732}{999} = $ _____ 15. $\dfrac{1957}{9999} = $ _____

Write each fractional equivalent.

16. $0.\overline{45} = $ _____ 17. $0.\overline{148} = $ _____ 18. $0.\overline{7213} = $ _____

19. Summarize your observations.

_____

_____

_____

Copyright © by Holt, Rinehart and Winston.
All rights reserved.

Holt Mathematics

Name _____ Date _____ Class _____

LESSON 2-1 Problem Solving
Rational Numbers

Write the correct answer.

1. Fill in the table below which shows the sizes of drill bits in a set.

2. Do the drill bit sizes convert to repeating or terminating decimals?

   _____ _____

13-Piece Drill Bit Set

Fraction	Decimal	Fraction	Decimal	Fraction	Decimal
$\frac{1}{4}"$		$\frac{11}{64}"$		$\frac{3}{32}"$	
$\frac{15}{64}"$		$\frac{5}{32}"$		$\frac{5}{64}"$	
$\frac{7}{32}"$		$\frac{9}{64}"$		$\frac{1}{16}"$	
$\frac{13}{64}"$		$\frac{1}{8}"$			
$\frac{3}{16}"$		$\frac{7}{64}"$			

Use the table at the right that lists the world's smallest nations. Choose the letter for the best answer.

3. What is the area of Vatican City expressed as a fraction in simplest form?

   A $\frac{8}{50}$ C $\frac{17}{1000}$

   B $\frac{4}{25}$ D $\frac{17}{100}$

World's Smallest Nations

Nation	Area (square miles)
Vatican City	0.17
Monaco	0.75
Nauru	8.2

4. What is the area of Monaco expressed as a fraction in simplest form?

   F $\frac{75}{100}$ H $\frac{3}{4}$

   G $\frac{15}{20}$ J $\frac{2}{3}$

5. What is the area of Nauru expressed as a mixed number?

   A $8\frac{1}{50}$ C $8\frac{2}{100}$

   B $8\frac{2}{50}$ D $8\frac{1}{5}$

6. The average annual precipitation in Miami, FL is 57.55 inches. Express 57.55 as a mixed number.

   F $57\frac{11}{20}$ H $57\frac{5}{100}$

   G $57\frac{55}{1000}$ J $57\frac{1}{20}$

7. The average annual precipitation in Norfolk, VA is 45.22 inches. Express 45.22 as a mixed number.

   A $45\frac{11}{50}$ C $45\frac{11}{20}$

   B $45\frac{22}{1000}$ D $45\frac{1}{5}$

Reading Strategies
Lesson 2-1: Use a Graphic Organizer

Definition
The set of numbers that can be written in the form $\frac{a}{b}$, where a and b are integers and b does not equal 0.

Facts
Fractions are rational numbers.
Decimals that terminate or repeat are rational numbers.
Whole numbers are rational numbers.
Integers are rational numbers.
0 is a rational number.

Rational Numbers

Examples
$2 = \frac{2}{1}$
$\frac{7}{8}$
$0.37 = \frac{37}{100}$
$4\frac{1}{4} = \frac{17}{4}$
$-5 = -\frac{5}{1}$

Non-examples
$\sqrt{3}$ (the square root of 3)
π (3.14159…)
$\sqrt{3}$ and π cannot be written as decimals that terminate or repeat.

Use the chart to answer the following questions.

1. What is a rational number?

2. Is 0.62 a rational number? Why or why not?

3. Is $2\frac{1}{3}$ a rational number? Why or why not?

4. Is $\sqrt{7}$ a rational number? Why or why not?

5. Is −8 a rational number? Why or why not?

6. Is 0 a rational number? Why or why not?

Name _____ Date _____ Class _____

Puzzles, Twisters & Teasers
LESSON 2-1 *Let's Be Rational!*

Circle words from the list in the word search. Then find a word that answers the riddle. Circle it and write it on the line.

rational	equivalent	numerator	denominator	relatively
prime	simplify	integer	nonzero	factor

```
H O L E Q U I V A L E N T N C
M I R U J I N O N Z E R O U Y
O P R F A C T O R X D E B M F
Q W A U I O E W E R T Y L E I
O T T L P J G O K I Y T R R L
B Y I M J U E W E R T Y U A P
N J O A S P R I M E T Y U T M
D E N O M I N A T O R I L O I
X W A R E L A T I V E L Y R S
P L L Q A Z X S W E D C V F R
```

What can you put in a bucket of water to make it lighter?

a _____ _____ _____ _____

Practice A
LESSON 2-2 Comparing and Ordering Rational Numbers

Compare. Write <, >, or =.

1. $\frac{3}{4}$ ☐ $\frac{5}{7}$

   $\frac{}{28}$ ☐ $\frac{}{28}$

   $\frac{3}{4}$ ☐ $\frac{5}{7}$

2. $-\frac{2}{5}$ ☐ $-\frac{3}{8}$

   $-\frac{}{40}$ ☐ $-\frac{}{40}$, so

   $-\frac{2}{5}$ ☐ $-\frac{3}{8}$

3. 0.3 ☐ $\frac{1}{4}$

   0.3 ☐ $\underline{}$

   0.3 ☐ $\frac{1}{4}$

4. $\frac{1}{6}$ ☐ $\frac{1}{3}$

   ☐
   $\underline{}$

   $\frac{1}{6}$ ☐ $\frac{1}{3}$

5. 0.09 ☐ $\frac{1}{2}$

   ☐
   $\underline{}$

   0.09 ☐ $\frac{1}{2}$

6. $-\frac{3}{5}$ ☐ -0.6

   ☐
   $\underline{}$

   $-\frac{3}{5}$ ☐ -0.6

7. $\frac{4}{5}$ ☐ $\frac{7}{10}$

8. $-\frac{1}{4}$ ☐ $-\frac{3}{4}$

9. $-\frac{2}{3}$ ☐ $-\frac{1}{4}$

10. $\frac{5}{8}$ ☐ $\frac{5}{6}$

11. $\frac{7}{9}$ ☐ $\frac{2}{3}$

12. $-\frac{3}{5}$ ☐ $-\frac{1}{10}$

13. $\frac{1}{5}$ ☐ $\frac{1}{8}$

14. $-1\frac{4}{5}$ ☐ -1.3

15. $1\frac{2}{3}$ ☐ $1\frac{2}{5}$

16. Trail A is $2\frac{2}{5}$ miles long. Trail B is $\frac{1}{4}$ mile long. Trail C is $1\frac{9}{10}$ miles long. Trail D is 2.05 miles long. List the lengths of the trails from shortest to longest.

_____

Name _____ Date _____ Class _____

LESSON 2-2 Practice B
Comparing and Ordering Rational Numbers

Compare. Write <, >, or =.

1. $\dfrac{1}{8}\ \square\ \dfrac{1}{10}$

2. $\dfrac{3}{5}\ \square\ \dfrac{7}{10}$

3. $-\dfrac{1}{3}\ \square\ -\dfrac{3}{4}$

4. $\dfrac{5}{6}\ \square\ \dfrac{3}{4}$

5. $-\dfrac{2}{7}\ \square\ -\dfrac{1}{2}$

6. $1\dfrac{2}{9}\ \square\ 1\dfrac{2}{3}$

7. $-\dfrac{8}{9}\ \square\ -\dfrac{3}{10}$

8. $-\dfrac{4}{5}\ \square\ -\dfrac{8}{10}$

9. $0.08\ \square\ \dfrac{3}{10}$

10. $\dfrac{11}{15}\ \square\ 0.7\overline{3}$

11. $2\dfrac{4}{9}\ \square\ 2\dfrac{3}{4}$

12. $-\dfrac{5}{8}\ \square\ -0.58$

13. $3\dfrac{1}{4}\ \square\ 3.3$

14. $-\dfrac{1}{6}\ \square\ -\dfrac{1}{9}$

15. $0.75\ \square\ \dfrac{3}{4}$

16. $-2\dfrac{1}{8}\ \square\ -2.1$

17. $1\dfrac{1}{2}\ \square\ 1.456$

18. $-\dfrac{3}{5}\ \square\ -0.6$

19. On Monday, Gina ran 1 mile in 9.3 minutes. Her times for running 1 mile on each of the next four days, relative to her time on Monday, were $-1\dfrac{2}{3}$ minutes, -1.45 minutes, -1.8 minutes, and $-1\dfrac{3}{8}$ minutes. List these relative times in order from least to greatest.

_____

20. Trail A is 3.1 miles long. Trail C is $3\dfrac{1}{4}$ miles long. Trail B is longer than Trail A but shorter than Trail C. What is a reasonable distance for the length of Trail B?

_____

Name _____ Date _____ Class _____

LESSON 2-2 Practice C
Comparing and Ordering Rational Numbers

Compare. Write <, >, or =.

1. $\dfrac{1}{9}\ \square\ \dfrac{1}{20}$ 2. $\dfrac{3}{8}\ \square\ \dfrac{5}{6}$ 3. $-\dfrac{1}{6}\ \square\ -\dfrac{2}{3}$

4. $\dfrac{11}{20}\ \square\ 0.55$ 5. $-\dfrac{4}{9}\ \square\ -\dfrac{1}{2}$ 6. $1\dfrac{3}{5}\ \square\ 1.35$

7. $-\dfrac{5}{9}\ \square\ -0.45$ 8. $-1\dfrac{7}{8}\ \square\ -1.875$ 9. $\dfrac{5}{4}\ \square\ 1.4$

10. $-1\dfrac{1}{5}\ \square\ -1.06$ 11. $4.00\ \square\ \dfrac{24}{4}$ 12. $-\dfrac{5}{12}\ \square\ -0.56$

Write a fraction or decimal that has a value between the given numbers.

13. $\dfrac{1}{6}$ and $\dfrac{1}{5}$ 14. 0.7 and 0.71   15. $-\dfrac{1}{6}$ and 0.2

_____   _____   _____

16. $-\dfrac{1}{2}$ and $-\dfrac{1}{4}$ 17. 1.45 and 1.46   18. $\dfrac{2}{3}$ and 0.75

_____   _____   _____

19. The students in one English class are reading the same book. Last night, James read $\dfrac{1}{4}$ of the book. Jennie read $\dfrac{3}{8}$ of the book. Kyle read 0.4 of the book, and Talia read 0.33 of the book. List the numbers in order from least to greatest. Who read the greatest number of pages last night?

_____

20. Melanie ran $2\dfrac{5}{8}$ miles on Monday. On Friday she ran 2.8 miles. On Wednesday, she ran further than on Monday but not as far as on Friday. What is a reasonable fraction length for the distance Melanie ran on Wednesday? What is a reasonable decimal length for the distance?

_____

Reteach
2-2 Comparing and Ordering Rational Numbers

You can use number lines to compare two fractions that have different denominators.

Compare $\frac{3}{8}$ and $\frac{2}{3}$.

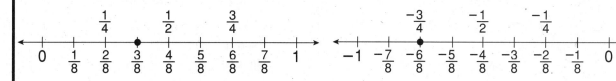

$\frac{3}{8} < \frac{2}{3}$

Compare $-\frac{3}{4}$ and $-\frac{5}{6}$.

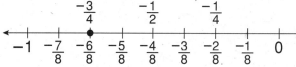

$-\frac{3}{4} > -\frac{5}{6}$

Use the number lines above. Write < or >.

1. $\frac{1}{4} \square \frac{1}{6}$
2. $\frac{5}{6} \square \frac{1}{2}$
3. $-\frac{2}{3} \square -\frac{1}{4}$
4. $-\frac{5}{6} \square -\frac{5}{8}$

You can also use number lines to compare a fraction and a decimal.

Compare 0.2 and $\frac{1}{3}$.

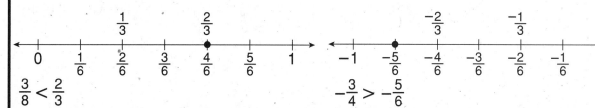

$0.2 < \frac{1}{3}$

Compare $-\frac{5}{8}$ and -0.9.

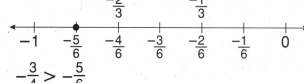

$-\frac{5}{8} > -0.9$

Use the number lines above. Write < or >.

5. $\frac{5}{6} \square 0.5$
6. $0.6 \square \frac{2}{3}$
7. $-0.4 \square -\frac{1}{4}$
8. $-\frac{7}{8} \square -0.8$

Name _____ Date _____ Class _____

LESSON 2-2 Challenge
From Repeating Decimal to Fraction

You can use an equation to write a repeating decimal as a fraction.

Write 0.3333… as a fraction.	Write 0.363636… as a fraction.
Let $x = 0.3333…$ Then $10x = 3.3333…$ $10x = 3.3333…$ $\underline{-x = 0.3333…}$ $9x = 3$ Subtract. $\dfrac{9x}{9} = \dfrac{3}{9}$ Solve the equation. $x = \dfrac{1}{3}$ Simplify. So, $0.3333… = \dfrac{1}{3}$.	Let $x = 0.363636…$ Then $100x = 36.363636…$ $100x = 36.3636…$ $\underline{-x = 0.3636…}$ $99x = 36$ Subtract. $\dfrac{99x}{99} = \dfrac{36}{99}$ Solve the equation. $x = \dfrac{4}{11}$ Simplify. So, $0.3636… = \dfrac{4}{11}$.

Write each repeating decimal as a fraction.

1. $0.6666… = $ _____

2. $0.8888… = $ _____

3. $0.454545… = $ _____

4. $0.090909… = $ _____

5. $0.636363… = $ _____

6. $0.16666… = $ _____

7. $0.2222… = $ _____

8. $0.83333… = $ _____

9. $0.41666… = $ _____

10. $0.58333… = $ _____

Name _____ Date _____ Class _____

Problem Solving
LESSON 2-2 Comparing and Ordering Rational Numbers

Write the correct answer.

1. Carl Lewis won the gold medal in the long jump in four consecutive Summer Olympic games. He jumped 8.54 meters in 1984, 8.72 meters in 1988, 8.67 meters in 1992, 8.5 meters in 1996. Order the length of his winning jumps from least to greatest.

   _____

2. Scientists aboard a submarine are gathering data at an elevation of $-42\frac{1}{2}$ feet. Scientists aboard a submersible are taking photographs at an elevation of $-45\frac{1}{3}$ feet. Which scientists are closer to the surface of the ocean?

   _____

3. The depth of a lake is measured at three different points. Point A is −15.8 meters, Point B is −17.3 meters, and Point C is −16.9 meters. Which point has the greatest depth?

   _____

4. At a swimming meet, Gail's time in her first heat was $42\frac{3}{8}$ seconds. Her time in the second heat was 42.25 seconds. Which heat did she swim faster?

   _____

The table shows the top times in a 5 K race. Choose the letter of the best answer.

5. Who had the fastest time in the race?
   A Marshall
   B Renzo
   C Dan
   D Aaron

Name	Time (minutes)
Marshall	18.09
Renzo	17.38
Dan	17.9
Aaron	18.61

6. Which is the slowest time in the table?
   F 18.09 minutes
   G 17.38
   H 17.9 minutes
   J 18.61 minutes

7. Aaron's time in a previous race was less than his time in this race but greater than Marshall's time in this race. How fast could Aaron have run in the previous race?
   A 19.24 min   C 18.35 min
   B 18.7 min   D 18.05 mi

LESSON 2-2 Reading Strategies
Identify Relationships

0, $\frac{1}{2}$, and 1 are common benchmarks for fractions. Sometimes you can use these benchmarks to compare fractions.

A fraction is close to 0 if its numerator is small compared to its denominator. Examples are $\frac{2}{11}$, $\frac{3}{20}$, and $\frac{4}{25}$.

A fraction is close to $\frac{1}{2}$ if its denominator is about twice as great as its denominator. Examples are $\frac{5}{11}$, $\frac{8}{15}$, and $\frac{10}{21}$.

A fraction is close to 1 if its numerator and denominator are close in value. Examples are $\frac{9}{10}$, $\frac{13}{15}$, and $\frac{17}{33}$.

To compare $\frac{7}{15}$ and $\frac{3}{22}$, look at the relationship between the numerator and denominator of each fraction.

15 is a little more than 2×7, so $\frac{1}{2}$ is a benchmark for $\frac{7}{15}$.

3 is much smaller than 22, so 0 is a benchmark for $\frac{3}{22}$.

Since $\frac{1}{2} > 0$, $\frac{7}{15} > \frac{3}{22}$.

Answer each question.

1. What is a benchmark for $\frac{35}{67}$? _____

2. What is a benchmark for $\frac{11}{13}$? _____

3. Use > or < to compare $\frac{35}{67}$ and $\frac{11}{13}$. _____

4. What is a benchmark for $\frac{17}{20}$? _____

5. What is a benchmark for $\frac{6}{35}$? _____

6. Use > or < to compare $\frac{17}{20}$ and $\frac{6}{35}$. _____

7. Use benchmarks to compare $\frac{21}{40}$ and $\frac{19}{21}$. Explain your thinking.

_____

Name _____ Date _____ Class _____

LESSON 2-2 Puzzles, Twisters, and Teasers
Rational Riddle

Why did Francis Fraction want to become a psychologist?

To answer the riddle, write the numbers in order from least to greatest.

Then write the corresponding letters in the same order.

$-1\frac{1}{4}$ A   _____ _____

-2.05 E   _____ _____

$-\frac{1}{5}$ T   _____ _____

$\frac{3}{4}$ N   _____ _____

0 I   _____ _____

-2.7 S   _____ _____

$-\frac{5}{6}$ R   _____ _____

$-1\frac{1}{8}$ S   _____ _____

0.95 L   _____ _____

-0.6 A   _____ _____

-1.3 W   _____ _____

0.8 A   _____ _____

$-2\frac{1}{2}$ H   _____ _____

$\frac{1}{8}$ O   _____ _____

Answer: _____

Name _____ Date _____ Class _____

LESSON 2-3 Practice A
Adding and Subtracting Rational Numbers

1. A statue $8\frac{5}{16}$ in. high rests on a stand that is $1\frac{3}{16}$ in. high. What is the total height?

   _____

2. During the 19th Olympic Winter Games in 2002, the United States 4-man bobsled teams won silver and bronze medals. USA-1 sled had a total time of 3 min 7.81 sec. The USA-2 sled had a total time of 3 min 7.86 sec. What is the difference in the time of the two runs?

   _____

Use a number line to find each sum.

3. $-0.2 + 0.6$

   _____

   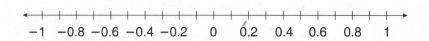

4. $\frac{1}{5} + \frac{3}{5}$

   _____

   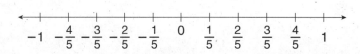

Add or subtract. Write each answer in simplest form.

5. $\frac{2}{9} + \frac{4}{9}$ 6. $\frac{5}{12} + \frac{3}{12}$ 7. $\frac{9}{10} - \frac{7}{10}$ 8. $\frac{8}{15} - \frac{11}{15}$

   _____   _____   _____   _____

9. $\frac{3}{14} - \frac{9}{14}$ 10. $\frac{5}{18} - \frac{11}{18}$ 11. $\frac{1}{8} + \frac{5}{8}$ 12. $\frac{5}{6} + \frac{1}{6}$

   _____   _____   _____   _____

Evaluate each expression for the given value of the variable.

13. $18.3 + x$ for $x = -1.6$ 14. $20.6 + x$ for $x = 2.8$ 15. $\frac{9}{11} + x$ for $x = -\frac{5}{11}$

   _____   _____   _____

Name _____ Date _____ Class _____

Practice B
LESSON 2-3 Adding and Subtracting Rational Numbers

1. Gretchen bought a sweater for $23.89. In addition, she had to pay $1.43 in sales tax. She gave the sales clerk $30. How much change did Gretchen receive from her total purchase?

   _____

2. Jacob is replacing the molding around two sides of a picture frame. The measurements of the sides of the frame are $4\frac{3}{16}$ in. and $2\frac{5}{16}$ in. What length of molding will Jacob need?

   _____

Use a number line to find each sum.

3. $-0.5 + 0.4$

   _____

   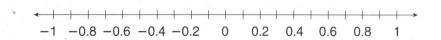

4. $-\frac{2}{7} + \frac{6}{7}$

   _____

   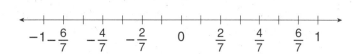

Add or subtract. Simplify.

5. $\frac{3}{8} + \frac{1}{8}$

6. $-\frac{1}{10} + \frac{7}{10}$

7. $\frac{5}{14} - \frac{3}{14}$

8. $\frac{4}{15} + \frac{7}{15}$

9. $\frac{5}{18} - \frac{7}{18}$

10. $-\frac{8}{17} - \frac{2}{17}$

11. $-\frac{1}{16} + \frac{5}{16}$

12. $\frac{3}{20} + \frac{1}{20}$

Evaluate each expression for the given value of the variable.

13. $38.1 + x$ for $x = -6.1$

14. $18.7 + x$ for $x = 8.5$

15. $\frac{8}{15} + x$ for $x = -\frac{4}{15}$

Copyright © by Holt, Rinehart and Winston.
All rights reserved.

Holt Mathematics

Name _____ Date _____ Class _____

LESSON 2-3 Practice C
Adding and Subtracting Rational Numbers

1. Jesse baked a pizza and cut it into 8 pieces. He ate three pieces and his two brothers ate two pieces each. In fractional form, how much of the pizza is left?

2. The biathlon combines cross-country skiing with rifle shooting at fixed targets. A biathlon competition features races of 6.2 mi with the contestants stopping to shoot twice, a race of 12.45 mi with four shooting stops, and a four person relay race totaling 18.6 mi. How many total miles are covered in a biathlon competition?

3. The home plate in baseball is a five-sided figure with sides that measure 17 in., 12 in., $8\frac{1}{2}$ in., 12 in., and $8\frac{1}{2}$ in. What is the sum of the sides of the figure?

4. Chato had $2032.64 in his checking account. He wrote two checks, one for $714.53 and the other for $289.67. What was the new balance in his checking account?

Add or subtract. Simplify.

5. $\frac{9}{17} + \frac{3}{17}$

6. $-\frac{4}{25} + \frac{19}{25}$

7. $\frac{17}{30} + \frac{7}{30}$

8. $-\frac{13}{20} + \frac{9}{20}$

9. $-\frac{8}{15} - \frac{4}{15}$

10. $\frac{5}{24} + \frac{13}{24}$

11. $\frac{13}{18} - \frac{7}{18}$

12. $-\frac{33}{50} - \frac{7}{50}$

Evaluate each expression for the given value of the variable.

13. $102.943 + x$ for $x = 2.03$

14. $\frac{12}{25} - x$ for $x = -\frac{13}{25}$

15. $18.01 - x$ for $x = -19.26$

Name _____ Date _____ Class _____

LESSON 2-3 Reteach
Adding and Subtracting Rational Numbers

To add fractions that have the same denominator:
- Use the common denominator for the sum.
- Add the numerators to get the numerator of the sum.
- Write the sum in simplest form.

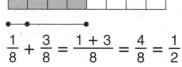

$$\frac{1}{8} + \frac{3}{8} = \frac{1+3}{8} = \frac{4}{8} = \frac{1}{2}$$

To subtract fractions that have the same denominator:
- Use the common denominator for the difference.
- Subtract the numerators.
    Subtraction is addition of an opposite.
- Write the difference in simplest form.

$$\frac{3}{6} - \left(-\frac{1}{6}\right) = \frac{3+1}{6} = \frac{4}{6} = \frac{2}{3}$$

Complete to add the fractions.

1. $\frac{3}{14} + \frac{4}{14} = $ _____ = _____

2. $\frac{2}{10} + \left(-\frac{4}{10}\right) = $ _____ = _____

3. $-\frac{5}{12} + \left(-\frac{3}{12}\right) = $ _____ = _____

Complete to subtract the fractions.

4. $\frac{8}{9} - \frac{2}{9} = $ _____ = _____

5. $\frac{9}{15} - \left(-\frac{3}{15}\right) = $ _____ = _____

6. $-\frac{10}{24} - \left(-\frac{2}{24}\right) = $ _____ = _____

To add or subtract decimals, line up the decimal points and then add or subtract from right to left as usual.

```
  12.83          35.78
 +24.17         -14.55
  37.00          21.23
```

Complete to add the decimals.

7. $14.23 + 3.56 = $ _____

8. $44.02 + 8.07 = $ _____

9. $1.39 + 13.6 = $ _____

Complete to subtract the decimals.

10. $124.33 - 13.16 = $ _____

11. $33.47 - 0.6 = $ _____

12. $25.15 - 25.06 = $ _____

Name _____ Date _____ Class _____

Lesson 2-3 Challenge
Number Code

Each sum is the code for a letter. As you find a sum, write its letter code in the message below. Write the sum in simplest form. Some letters appear more than once. An example is done for you.

$4.5 + (-6.5)$ ___-2___, T 1. $14.56 + (-10.09)$ _____, V

2. $\frac{7}{8} + \left(-1\frac{3}{8}\right)$ _____, M 3. $\frac{6}{8} + \left(-\frac{3}{8}\right)$ _____, N

4. $-1.05 + 0.85$ _____, I 5. $\frac{-2}{4} + \left(\frac{-3}{4}\right)$ _____, U

6. $-7.08 + (-12.02)$ _____, S 7. $-9.5 + 3.1$ _____, E

8. $\frac{-4}{5} + 1$ _____, E 9. $-1\frac{1}{2} + \left(-1\frac{1}{2}\right)$ _____, H

10. $1\frac{2}{4} + \left(\frac{-3}{4}\right)$ _____, P 11. $5 + \left(-4\frac{1}{10}\right)$ _____, I

12. $8 + (-6.4)$ _____, Y 13. $-3\frac{1}{4} + 3\frac{1}{4}$ _____, S

14. $\frac{7}{8} + \left(-1\frac{7}{8}\right)$ _____, L 15. $6.52 + (-5)$ _____, Z

16. $-62.3 + 23.9$ _____, A 17. $9\frac{1}{8} + (-10)$ _____, R

18. $-2.9 + 0.85$ _____, O 19. $2.7 + (-0.9)$ _____, O

$\underline{}\;\underline{}\;\underline{}\;\underline{}\qquad\underline{}\;\underline{}\;\underline{}\qquad\underline{}\;\underline{}\;\underline{}$
$-19.1\;\;-1\frac{1}{4}\;\;-\frac{7}{8}\;\;-6.4\qquad 1.6\;\;-2.05\;\;-1\frac{1}{4}\qquad -38.4\;\;-\frac{7}{8}\;\;\frac{1}{5}$

$\underline{}\;\underline{}\;\underline{\text{T}}\qquad\underline{}\;\underline{}\;\underline{}\;\underline{}\qquad\underline{\text{T}}\;\underline{}\;\underline{}\;\underline{}$
$\frac{3}{8}\;\;1.8\;\;-2\qquad -1\;\;\frac{1}{5}\;\;-19.1\;\;0\qquad -2\;\;-3\;\;-38.4\;\;\frac{3}{8}$

$\underline{}\;\underline{}\;\underline{}\;\underline{}\;?$
$1.52\;\;-6.4\;\;-\frac{7}{8}\;\;-2.05$

$\underline{}\;\underline{}\qquad\underline{}\;\underline{}\;\underline{}\qquad\underline{}\;\underline{\text{T}}\;\underline{}\;\underline{}\;\underline{}\;!$
$-0.2\;\;-\frac{1}{2}\qquad \frac{3}{4}\;\;-2.05\;\;0\qquad \frac{9}{10}\;\;-2\;\;-0.2\;\;4.47\;\;\frac{1}{5}$

Name _____ Date _____ Class _____

Problem Solving
LESSON 2-3 Adding and Subtracting Rational Numbers

Write the correct answer.

1. In 2004, Yuliya Nesterenko of Belarus won the Olympic Gold in the 100-m dash with a time of 10.93 seconds. In 2000, American Marion Jones won the 100-m dash with a time of 10.75 seconds. How many seconds faster did Marion Jones run the 100-m dash?

2. The snowfall in Rochester, NY in the winter of 1999–2000 was 91.5 inches. Normal snowfall is about 76 inches per winter. How much more snow fell in the winter of 1999–2000 than is normal?

3. In a survey, $\frac{76}{100}$ people indicated that they check their e-mail daily, while $\frac{23}{100}$ check their e-mail weekly, and $\frac{1}{100}$ check their e-mail less than once a week. What fraction of people check their e-mail at least once a week?

4. To make a small amount of play dough, you can mix the following ingredients: 1 cup of flour, $\frac{1}{2}$ cup of salt and $\frac{1}{2}$ cup of water. What is the total amount of ingredients added to make the play dough?

Choose the letter for the best answer.

5. How much more expensive is it to buy a ticket in Boston than in Minnesota?
   A $20.95
   B $55.19
   C $5.40
   D $26.35

Baseball Ticket Prices

Location	Average Price
Minnesota	$14.42
League Average	$19.82
Boston	$40.77

6. How much more expensive is it to buy a ticket in Boston than the league average?
   F $60.59
   G $20.95
   H $5.40
   J $26.35

7. What is the total cost of a ticket in Boston and a ticket in Minnesota?
   A $55.19
   B $34.24
   C $60.59
   D $54.19

Name _____ Date _____ Class _____

LESSON 2-3 Reading Strategies
Use a Visual Model

A number line can help you picture addition with decimals. This number line is divided into tenths.

Add $-0.6 + 1.8$.

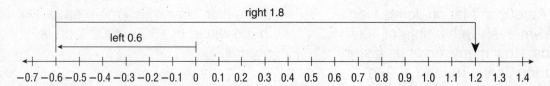

1. Where do you start on the number line? _____

2. Do you move to the right or to the left? Why?
   _____

3. How many places do you move? _____

4. To add, do you move to the right or to the left? _____

5. How many places do you move? _____

6. At what number do you end? _____

This number line helps you picture addition with fractions. The number line is divided into sixths.

Add $-\frac{4}{6} + 1\frac{3}{6}$.

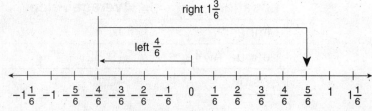

7. Do you move first to the left or right from 0 on the number line? Why?
   _____

8. How many places do you move? _____

9. To add, do you move to the right or left? _____

10. How many places to you move? _____

11. At what number do you end? _____

Name _____ Date _____ Class _____

LESSON 2-3 Puzzles, Twisters & Teasers
Bee a Math Master!

Add or subtract to find the answers. Then solve the riddle using the letters associated with the answers.

S $\quad -\dfrac{1}{12} + \left(-\dfrac{7}{12}\right) =$ _____

F $\quad -0.9 + 2.5 =$ _____

D $\quad \dfrac{8}{11} - \dfrac{3}{11} =$ _____

O $\quad -\dfrac{4}{13} - \dfrac{8}{13} =$ _____

R $\quad -\dfrac{1}{15} + \dfrac{13}{15} =$ _____

G $\quad \dfrac{11}{32} - \dfrac{27}{32} =$ _____

W $\quad -0.06 + 0.86 =$ _____

T $\quad -\dfrac{19}{25} + \dfrac{13}{25} =$ _____

E $\quad \dfrac{8}{21} + \dfrac{15}{21} =$ _____

H $\quad 0.9 + 0.3 =$ _____

Why did the bee hum?

It $\underset{1.6}{\rule{1cm}{0.4pt}} \underset{-\frac{12}{13}}{\rule{1cm}{0.4pt}} \underset{\frac{4}{5}}{\rule{1cm}{0.4pt}} \underset{-\frac{1}{2}}{\rule{1cm}{0.4pt}} \underset{-\frac{12}{13}}{\rule{1cm}{0.4pt}} \underset{-\frac{6}{25}}{\rule{1cm}{0.4pt}} \qquad \underset{-\frac{6}{25}}{\rule{1cm}{0.4pt}} \underset{1.2}{\rule{1cm}{0.4pt}} \underset{\frac{23}{21}}{\rule{1cm}{0.4pt}}$

$\underset{0.8}{\rule{1cm}{0.4pt}} \underset{-\frac{12}{13}}{\rule{1cm}{0.4pt}} \underset{\frac{4}{5}}{\rule{1cm}{0.4pt}} \underset{\frac{5}{11}}{\rule{1cm}{0.4pt}} \underset{-\frac{2}{3}}{\rule{1cm}{0.4pt}}$.

Name _____ Date _____ Class _____

LESSON 2-4 Practice A
Multiplying Rational Numbers

Multiply. Write each answer in simplest form.

1. $5\left(\dfrac{1}{3}\right)$ 2. $-2\left(\dfrac{2}{5}\right)$ 3. $4\left(\dfrac{1}{6}\right)$ 4. $-3\left(\dfrac{2}{9}\right)$

5. $-\dfrac{5}{7}\left(\dfrac{2}{5}\right)$ 6. $\dfrac{3}{4}\left(\dfrac{1}{3}\right)$ 7. $-\dfrac{1}{4}\left(\dfrac{1}{3}\right)$ 8. $-\dfrac{1}{6}\left(-\dfrac{2}{3}\right)$

9. $\dfrac{1}{2}\left(\dfrac{10}{7}\right)$ 10. $\dfrac{3}{10}\left(-\dfrac{5}{18}\right)$ 11. $\dfrac{4}{5}\left(-\dfrac{12}{16}\right)$ 12. $\dfrac{4}{3}\left(\dfrac{24}{16}\right)$

13. $-4\left(1\dfrac{1}{2}\right)$ 14. $\dfrac{3}{4}\left(\dfrac{5}{8}\right)$ 15. $-\dfrac{2}{5}\left(3\dfrac{1}{4}\right)$ 16. $-\dfrac{5}{6}\left(-\dfrac{3}{10}\right)$

Multiply.

17. -3.2×5 18. 0.34×0.06 19. -8.12×-9 20. 4.24×3.5

21. -3.14×0.007 22. -6.7×0.8 23. -0.25×-2.4 24. 7.9×-2

25. Jade babysat $4\dfrac{1}{2}$ hours for the Lenox family. She was paid $5 an hour. How much did she receive for this babysitting job?

_____

Copyright © by Holt, Rinehart and Winston.
All rights reserved.

Holt Mathematics

Practice B
2-4 Multiplying Rational Numbers

Multiply. Write each answer in simplest form.

1. $8\left(\dfrac{3}{4}\right)$
2. $-6\left(\dfrac{9}{18}\right)$
3. $-9\left(\dfrac{5}{6}\right)$
4. $-6\left(-\dfrac{7}{12}\right)$

5. $-\dfrac{5}{18}\left(\dfrac{8}{15}\right)$
6. $\dfrac{7}{12}\left(\dfrac{14}{21}\right)$
7. $-\dfrac{1}{9}\left(\dfrac{27}{24}\right)$
8. $-\dfrac{1}{11}\left(-\dfrac{3}{2}\right)$

9. $\dfrac{7}{20}\left(-\dfrac{15}{28}\right)$
10. $\dfrac{16}{25}\left(-\dfrac{18}{32}\right)$
11. $\dfrac{1}{9}\left(-\dfrac{18}{17}\right)$
12. $\dfrac{17}{20}\left(-\dfrac{12}{34}\right)$

13. $-4\left(2\dfrac{1}{6}\right)$
14. $\dfrac{3}{4}\left(1\dfrac{3}{8}\right)$
15. $3\dfrac{1}{5}\left(\dfrac{2}{3}\right)$
16. $-\dfrac{5}{6}\left(2\dfrac{1}{2}\right)$

Multiply.

17. $-2(-5.2)$
18. $0.53(0.04)$
19. $(-7)(-3.9)$
20. $-2(8.13)$

21. $0.02(-4.62)$
22. $0.5(-7.8)$
23. $(-0.41)(-8.5)$
24. $(-8)(6.3)$

25. $15(-0.05)$
26. $(-3.04)(-1.7)$
27. $10(-0.09)$
28. $(-0.8)(-0.15)$

29. Travis painted for $6\dfrac{2}{3}$ hours. He received $27 an hour for his work. How much was Travis paid for doing this painting job?

Name _____ Date _____ Class _____

LESSON 2-4 Practice C
Multiplying Rational Numbers

Multiply. Write each answer in simplest form.

1. $10\left(\dfrac{4}{5}\right)$

2. $-12\left(\dfrac{7}{24}\right)$

3. $-11\left(\dfrac{5}{22}\right)$

4. $-18\left(\dfrac{5}{36}\right)$

_____ _____ _____ _____

5. $\dfrac{14}{28}\left(\dfrac{7}{42}\right)$

6. $\dfrac{25}{64}\left(\dfrac{16}{75}\right)$

7. $-\dfrac{14}{19}\left(\dfrac{38}{70}\right)$

8. $-\dfrac{5}{27}\left(-\dfrac{9}{35}\right)$

_____ _____ _____ _____

9. $\dfrac{9}{20}\left(\dfrac{36}{81}\right)$

10. $1\dfrac{7}{10}\left(-\dfrac{5}{17}\right)$

11. $\dfrac{39}{50}\left(-\dfrac{35}{117}\right)$

12. $1\dfrac{1}{3}\left(-\dfrac{63}{168}\right)$

_____ _____ _____ _____

13. $-3\left(2\dfrac{32}{64}\right)$

14. $\dfrac{2}{38}\left(9\dfrac{1}{2}\right)$

15. $-\dfrac{16}{24}\left(1\dfrac{24}{32}\right)$

16. $3\dfrac{1}{4}\left(\dfrac{8}{48}\right)$

_____ _____ _____ _____

Multiply.

17. $15(-13.5)$

18. $6.34(1.08)$

19. $(-19)(-11.82)$

20. $8.5(16.42)$

_____ _____ _____ _____

21. $3.08(-4.38)$

22. $2.8(7.15)$

23. $(-2.25)(-2.25)$

24. $(16)(2.001)$

_____ _____ _____ _____

25. Mrs. Johnson harvested 107 pounds of tomatoes from her garden. She sold them for $0.85 a pound. How much did she receive from selling all the tomatoes?

_____

26. A store is having a clearance sale of $\dfrac{1}{3}$ off the regular price. How much will be saved on a jacket with a regular price of $154.35? What will be the sale price of the jacket?

_____

Name _____ Date _____ Class _____

LESSON 2-4 Reteach
Multiplying Rational Numbers

To model $\frac{1}{3} \times \frac{3}{4}$:

Divide a square into 4 equal parts. Lightly shade 3 of the 4.	Darken 1 of the 3 shaded parts.	Compare the 1 darkened part to the original 4.
		$\frac{1}{3} \times \frac{3}{4} = \frac{1}{4}$

Model each multiplication. Write the result.

1.
$\frac{1}{2} \times \frac{2}{4} =$ _____

2.
$\frac{3}{4} \times \frac{4}{6} =$ _____

3.
$\frac{2}{3} \times \frac{3}{9} =$ _____

To multiply fractions:
- Cancel common factors, one in a numerator and the other in a denominator.
- Multiply the remaining factors in the numerator and in the denominator.
- If the signs of the factors are the same, the product is positive. If the signs of the factors are different, the product is negative.

$\frac{\cancel{3}^1}{\cancel{4}_1} \times \frac{\cancel{8}^2}{\cancel{9}_3} = \frac{1 \times 2}{1 \times 3} = \frac{2}{3}$

Multiply. Answer in simplest form.

4. $\frac{1}{2} \times \frac{4}{9} =$ _____

5. $\frac{2}{3} \times \frac{6}{7} =$ _____

6. $\frac{3}{5} \times \frac{15}{17} =$ _____

7. $\frac{2}{3} \times \left(-\frac{9}{10}\right) =$ _____

8. $\left(-\frac{2}{9}\right) \times \frac{27}{40} =$ _____

9. $\left(-\frac{4}{7}\right) \times \left(-\frac{21}{8}\right) =$ _____

Name _____ Date _____ Class _____

LESSON 2-4 Challenge
Curtains

Variations of modern long multiplication were introduced into Europe by a 13th-century Italian, Leonardo of Pisa (Fibonacci). Many multiplication techniques can be traced to a book called *Lilaviti*, written by Bhaskara for his daughter in 12th-century India.

Here's how to do multiplication by the **Gelosia Method**, named after *jalousie*, the iron grill Italians placed over the windows. The method is also called the **Lattice Method of Multiplication**.

Consider: 38 × 56

Divide a square as shown.

Align the factors. Insert the individual products.

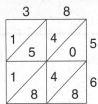

Sum each diagonal; begin with lower right. As needed, carry numbers into next diagonal sum.

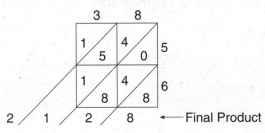

So, 38 × 56 = 2128.

Use the Gelosia Method to multiply. Verify results by your usual method.

1.

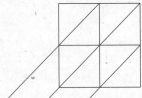

39 × 47 = _____

2.

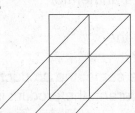

68 × 73 = _____

3.

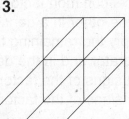

358 × 64 = _____

Name _____ Date _____ Class _____

LESSON 2-4 Problem Solving
Multiplying Rational Numbers

Use the table at the right.

1. What was the average number of births per minute in 2001?

   _____

Average World Births and Deaths per Second in 2001

Births	$4\frac{1}{5}$
Deaths	1.7

2. What was the average number of deaths per hour in 2001?

   _____

3. What was the average number of births per day in 2001?

   _____

4. What was the average number of births in $\frac{1}{2}$ of a second in 2001?

   _____

5. What was the average number of births in $\frac{1}{4}$ of a second in 2001?

   _____

Use the table below. During exercise, the target heart rate is 0.5–0.75 of the maximum heart rate. Choose the letter for the best answer.

6. What is the target heart rate range for a 14 year old?

   A 7–10.5
   B 103–154.5
   C 145–166
   D 206–255

Age	Maximum Heart Rate
13	207
14	206
15	205
20	200
25	195

Source: American Heart Association

7. What is the target heart rate range for a 20 year old?

   F 100–150
   G 125–175
   H 150–200
   J 200–250

8. What is the target heart rate range for a 25 year old?

   A 25–75
   B 85–125
   C 97.5–146.25
   D 195–250

Name _____ Date _____ Class _____

LESSON 2-4
Reading Strategies
Use a Visual Model

This rectangle will help you understand how to find the product of $\frac{1}{2} \cdot \frac{1}{3}$. First, $\frac{1}{2}$ of the rectangle was shaded. Then, the rectangle was divided horizontally into thirds. Then, $\frac{1}{3}$ was shaded. The overlap of the shading shows the product of $\frac{1}{2} \cdot \frac{1}{3}$.

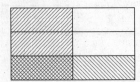

1. Into how many parts is the rectangle divided? What fractional part of the rectangle is each of these parts? _____

2. What fractional part of the rectangle has shading that overlaps? _____

3. Multiply the numerators and the denominators of the given fractions. _____ = _____

4. Use the rectangle to draw a model for the problem $\frac{1}{4} \cdot \frac{1}{2}$.

5. Draw lines from top to bottom to divide the rectangle into fourths. Shade one-fourth of the rectangle.

6. Draw a line across the rectangle to divide it into halves. Into how many parts is the rectangle now divided? _____

7. Shade one of the halves.

8. What fractional part of the rectangle was shaded twice? _____

9. Multiply the numerators and denominators. _____

Copyright © by Holt, Rinehart and Winston.
All rights reserved.

Holt Mathematics

Name _____ Date _____ Class _____

LESSON 2-4 Puzzles, Twisters & Teasers
Egg-zactly Correct!

Multiply. Write each answer in its simplest form. Then solve the riddle using the answers.

Y $\dfrac{7}{8}\left(\dfrac{3}{5}\right) =$ _____

N $3\left(2\dfrac{1}{5}\right) =$ _____

T $-2\left(\dfrac{9}{16}\right) =$ _____

E $6\left(\dfrac{2}{3}\right) =$ _____

I $-5\left(1\dfrac{3}{4}\right) =$ _____

W $2\left(\dfrac{7}{8}\right) =$ _____

C $\dfrac{3}{4}\left(-\dfrac{1}{8}\right) =$ _____

O $\dfrac{6}{8}\left(\dfrac{2}{5}\right) =$ _____

K $1\dfrac{2}{3}\left(\dfrac{5}{6}\right) =$ _____

L $-\dfrac{1}{3}\left(-\dfrac{4}{7}\right) =$ _____

What do you call a city with a million eggs?

$\overline{}\;\overline{}\;\overline{}\overline{}\;\overline{}\;\overline{}\;\overline{}$
$6\dfrac{3}{5}\quad 4\quad 1\dfrac{3}{4}\qquad \dfrac{21}{40}\quad \dfrac{3}{10}\quad \dfrac{4}{21}\quad 1\dfrac{7}{18}$

$\overline{}\;\overline{}\;\overline{}\;\overline{}$
$-\dfrac{3}{32}\quad -8\dfrac{3}{4}\quad -1\dfrac{1}{8}\quad \dfrac{21}{40}$

Name _____ Date _____ Class _____

LESSON 2-5

Practice A
Dividing Rational Numbers

Divide. Write each answer in simplest form.

1. $\dfrac{1}{4} \div \dfrac{3}{8}$
2. $-\dfrac{2}{3} \div \dfrac{5}{9}$
3. $\dfrac{1}{6} \div \dfrac{1}{3}$
4. $\dfrac{3}{4} \div \left(-\dfrac{1}{8}\right)$

5. $\dfrac{1}{9} \div \dfrac{1}{3}$
6. $\dfrac{2}{5} \div \dfrac{4}{7}$
7. $-\dfrac{3}{5} \div \dfrac{6}{7}$
8. $-\dfrac{3}{8} \div \left(-\dfrac{5}{6}\right)$

9. $1\dfrac{2}{5} \div 1\dfrac{1}{2}$
10. $-\dfrac{3}{4} \div 9$
11. $-2\dfrac{1}{3} \div \dfrac{1}{4}$
12. $-\dfrac{5}{8} \div 5$

Divide.

13. $1.53 \div 0.3$
14. $5.14 \div 0.2$
15. $10.05 \div 0.05$
16. $5.28 \div 0.4$

17. $6.54 \div 0.03$
18. $29.45 \div 0.005$
19. $8.58 \div 0.06$
20. $1.61 \div 0.7$

Evaluate each expression for the given value of the variable.

21. $\dfrac{10}{x}$ for $x = 0.05$
22. $\dfrac{9.12}{x}$ for $x = -0.2$
23. $\dfrac{42.42}{x}$ for $x = 1.4$

24. Mr. Chen has a 76-in. space to stack books. Each book is $6\dfrac{1}{3}$ in. tall. How many books can he stack in the space?

Name _____ Date _____ Class _____

LESSON 2-5 Practice B
Dividing Rational Numbers

Divide. Write each answer in simplest form.

1. $\dfrac{1}{5} \div \dfrac{3}{10}$

2. $-\dfrac{5}{8} \div \dfrac{3}{4}$

3. $\dfrac{1}{4} \div \dfrac{1}{8}$

4. $-\dfrac{2}{3} \div \dfrac{4}{15}$

5. $1\dfrac{2}{9} \div 1\dfrac{2}{3}$

6. $-\dfrac{7}{10} \div \left(\dfrac{2}{5}\right)$

7. $\dfrac{6}{11} \div \dfrac{3}{22}$

8. $\dfrac{4}{9} \div \left(-\dfrac{8}{15}\right)$

9. $\dfrac{3}{8} \div -15$

10. $-\dfrac{5}{6} \div 12$

11. $6\dfrac{1}{2} \div 1\dfrac{5}{8}$

12. $-\dfrac{9}{10} \div 6$

Divide.

13. $24.35 \div 0.5$

14. $2.16 \div 0.04$

15. $3.16 \div 0.02$

16. $7.32 \div 0.3$

17. $87.36 \div 0.6$

18. $79.36 \div 0.8$

19. $4.27 \div 0.007$

20. $63.81 \div 0.9$

21. $1.23 \div 0.003$

22. $62.46 \div 0.09$

23. $21.12 \div 0.4$

24. $82.68 \div 0.06$

Evaluate each expression for the given value of the variable.

25. $\dfrac{18}{x}$ for $x = 0.12$

26. $\dfrac{10.8}{x}$ for $x = 0.03$

27. $\dfrac{9.18}{x}$ for $x = -1.2$

28. A can of fruit contains $3\dfrac{1}{2}$ cups of fruit. The suggested serving size is $\dfrac{1}{2}$ cup. How many servings are in the can of fruit?

Name _____ Date _____ Class _____

LESSON 2-5
Practice C
Dividing Rational Numbers

Divide. Write each answer in simplest form.

1. $\dfrac{10}{15} \div \dfrac{8}{25}$

2. $1\dfrac{3}{18} \div 2\dfrac{1}{3}$

3. $\dfrac{6}{13} \div \dfrac{18}{26}$

4. $-\dfrac{12}{21} \div \dfrac{3}{14}$

5. $-\dfrac{3}{20} \div \left(\dfrac{15}{35}\right)$

6. $-\dfrac{27}{32} \div 3$

7. $\dfrac{36}{41} \div \dfrac{9}{82}$

8. $2\dfrac{12}{34} \div 1\dfrac{3}{17}$

Divide.

9. $3.15 \div 0.05$

10. $26.008 \div 0.4$

11. $-983.1 \div (-0.3)$

12. $1.44 \div 0.16$

13. $236.4 \div 0.0012$

14. $-10.08 \div 0.005$

15. $2.253 \div 0.15$

16. $-1.161 \div 0.18$

Evaluate each expression for the given value of the variable.

17. $\dfrac{2.25}{x}$ for $x = -0.009$

18. $\dfrac{-234.72}{x}$ for $x = 3.6$

19. $\dfrac{-4330.8}{x}$ for $x = -8.02$

20. Emma bought $2\dfrac{1}{2}$ yards of cording for the trim around the edge of a square pillow. How much will she use for each side of the pillow?

21. Sean has a loan of $8804.46 including interest. He makes payments of $209.63 each month on the simple interest loan. How many months will it take for Sean to repay his loan?

Holt Mathematics

Name _____ Date _____ Class _____

LESSON 2-5 Reteach
Dividing Rational Numbers

To write the **reciprocal** of a fraction, interchange the numerator and denominator.

The product of a number and its reciprocal is 1.

$$\frac{2}{3} \times \frac{3}{2} = 1$$

$\frac{2}{3} \rightleftarrows \frac{3}{2}$
Fraction   Reciprocal

Write the reciprocal of each rational number.

1. The reciprocal of $\frac{3}{5}$ is:

2. The reciprocal of 6 is:

3. The reciprocal of $2\frac{1}{3}$ is:

_____   _____   _____

To divide by a fraction, multiply by its reciprocal.

$\frac{2}{3} \div 6$

$\frac{2}{3} \times \frac{1}{6}$

$\frac{1 \times 1}{3 \times 3} = \frac{1}{9}$

$\frac{3}{5} \div \frac{9}{10}$

$\frac{3}{5} \times \frac{10}{9}$

$\frac{\overset{1}{3} \times \overset{2}{10}}{\underset{1}{5} \times \underset{3}{9}} = \frac{2}{3}$

Complete to divide and simplify.

4. $\frac{3}{8} \div 12 = \frac{3}{8} \times$ _____ = _____

5. $\frac{4}{3} \div 16 =$ _____

6. $\frac{5}{7} \div \frac{20}{21} = \frac{5}{7} \times$ _____ = _____

7. $-\frac{3}{4} \div \left(\frac{9}{8}\right) = -\frac{3}{4} \times$ _____ = _____

Change a decimal divisor to a whole number. Using the number of places in the divisor, move the decimal point to the right in both the divisor and the dividend.

$0.7\overline{)4.34} \rightarrow 0.7\overline{)4.3.4} \rightarrow 7\overline{)43.4}^{\,6.2}$

Rewrite each division with a whole-number divisor. Then, do the division.

8. $0.6\overline{)1.14} \rightarrow$ _____ = _____

9. $0.3\overline{)4.56} \rightarrow$ _____ = _____

10. $0.02\overline{)7.12} \rightarrow$ _____ = _____

11. $0.08\overline{)57.28} \rightarrow$ _____ = _____

Copyright © by Holt, Rinehart and Winston.
All rights reserved.

Holt Mathematics

Name _____ Date _____ Class _____

LESSON 2-5 Challenge
A New License to Operate

You can invent new operations based on the familiar operations of addition, subtraction, multiplication, and division, and the familiar order of operations.

If $a \triangle b = \dfrac{a+b}{2}$ where a and b represent any rational numbers,

then $3 \triangle 5 = \dfrac{3+5}{2} = 4$.

Use the given definition of operation $\triangle$ to evaluate each expression.

1. $\dfrac{1}{2} \triangle (-10) =$ _____

2. $\dfrac{100 \triangle (-10)}{10} =$ _____

3. $4 \triangle 6 \triangle 3 =$ _____

4. $[5.5 \triangle (-6)] + [-6 \triangle 5.5] =$ _____

Use the operation shown to answer each question. $\begin{array}{|c|c|}\hline a & b \\\hline c & d \\\hline\end{array} = ac - bd$

5. $\begin{array}{|c|c|}\hline 1 & 8 \\\hline 3 & 4 \\\hline\end{array} =$ _____

6. $\begin{array}{|c|c|}\hline -2 & 3 \\\hline 3 & -2 \\\hline\end{array} =$ _____

7. If $\begin{array}{|c|c|}\hline 1 & 3 \\\hline x & 2 \\\hline\end{array} = 18$, then $x =$ _____

8. If $\begin{array}{|c|c|}\hline 6 & 2 \\\hline x & x \\\hline\end{array} = 12$, then $x =$ _____

Use the operation shown to answer each question. $a \searrow b = \dfrac{a^2}{b^2}$

9. $(3 \searrow 5) =$ _____

10. $(1 \searrow 8) - (5 \searrow 8) =$ _____

11. $(1 \searrow 3) \times (3 \searrow 6) =$ _____

12. $(1 \searrow 10)^2 =$ _____

If $\lfloor n \rfloor$ means 1 less than the number of digits in the integer n, then, for example, $\lfloor 77 \rfloor = 1$ since 77 has 2 digits.

Use the definition of $\lfloor n \rfloor$ to answer each question.

13. If n is a positive integer less than 100, what is the greatest value for $\lfloor n \rfloor$? _____

14. If n is a positive integer less than 1001, what is the greatest value for $\lfloor n \rfloor$? _____

15. If n has 100 digits, what is the value of $\lfloor \lfloor n \rfloor \rfloor$? Explain.

_____

_____

Problem Solving
2-5 Dividing Rational Numbers

Use the table at the right that shows the maximum speed over a quarter mile of different animals. Find the time is takes each animal to travel one-quarter mile at top speed. Round to the nearest thousandth.

1. Quarter horse

2. Greyhound

3. Human

Maximum Speeds of Animals

Animal	Speed (mph)
Quarter Horse	47.50
Greyhound	39.35
Human	27.89
Giant Tortoise	0.17
Three-toed sloth	0.15

4. Giant tortoise

5. Three-toed sloth

Choose the letter for the best answer.

6. A piece of ribbon is $1\frac{7}{8}$ inches long. If the ribbon is going to be divided into 15 pieces, how long should each piece be?

   A $\frac{1}{8}$ in.
   B $\frac{1}{15}$ in.
   C $\frac{2}{3}$ in.
   D $28\frac{1}{8}$ in.

7. The recorded rainfall for each day of a week was 0 in., $\frac{1}{4}$ in., $\frac{3}{4}$ in., 1 in., 0 in., $1\frac{1}{4}$ in., $1\frac{1}{4}$ in. What was the average rainfall per day?

   F $\frac{9}{10}$ in.
   G $\frac{9}{14}$ in.
   H $\frac{7}{8}$ in.
   J $4\frac{1}{2}$ in.

8. A drill bit that is $\frac{7}{32}$ in. means that the hole the bit makes has a diameter of $\frac{7}{32}$ in. Since the radius is half of the diameter, what is the radius of a hole drilled by a $\frac{7}{32}$ in. bit?

   A $\frac{14}{32}$ in.   C $\frac{9}{16}$ in.
   B $\frac{7}{32}$ in.   D $\frac{7}{64}$ in.

9. A serving of a certain kind of cereal is $\frac{2}{3}$ cup. There are 12 cups of cereal in the box. How many servings of cereal are in the box?

   F 18
   G 15
   H 8
   J 6

Name _____ Date _____ Class _____

LESSON 2-5 Reading Strategies
Focus on Vocabulary

The word **reciprocal** means an exchange. When two friends exchange gifts, you might think of the gifts as "switching places." In the reciprocal of a fraction, the numerator and denominator exchange places.

Fraction	Reciprocal
$\frac{2}{3}$	$\frac{3}{2}$
$\frac{4}{5}$	$\frac{5}{4}$
$\frac{8}{1}$	$\frac{1}{8}$

1. What does the word *reciprocal* mean? _____

2. What is the reciprocal of $\frac{7}{8}$? _____

3. What is the reciprocal of $\frac{6}{5}$? _____

The product of a fraction and its reciprocal is always 1.

Fraction • Reciprocal = Product

$\frac{2}{3} \cdot \frac{3}{2} = \frac{6}{6} = 1$

$\frac{4}{5} \cdot \frac{5}{4} = \frac{20}{20} = 1$

$\frac{1}{8} \cdot \frac{8}{1} = \frac{8}{8} = 1$

4. What is the product of $\frac{1}{7} \cdot \frac{7}{1}$? _____

5. What is the product of $\frac{2}{6}$ and its reciprocal? _____

6. What is the reciprocal of $\frac{1}{2}$? _____

7. What is the product of $\frac{1}{2} \times 2$? _____

Copyright © by Holt, Rinehart and Winston.
All rights reserved.

Holt Mathematics

Puzzles, Twisters & Teasers

LESSON 2-5 *Hungry for Knowledge!*

Solve each equation. Round answers to the nearest tenth. Then use the answers to solve the riddle.

R $17.78 \div 0.7 =$ _____

L $\frac{1}{2} \div \frac{1}{4} =$ _____

E $10.08 \div 0.6 =$ _____

U $24 \div \frac{3}{4} =$ _____

N $3.72 \div 0.3 =$ _____

C $10\frac{1}{2} \div 3\frac{1}{2} =$ _____

I $14.08 \div 0.8 =$ _____

H $3\frac{1}{3} \div 1\frac{1}{5} =$ _____

D $9.36 \div 0.03 =$ _____

A $6.52 \div 0.004 =$ _____

What are two things you can't eat for breakfast?

__ __ __ __ __   __ __ __
2   32 12.4 3 2.8   1630 12.4 312

__ __ __ __ __ __
312 17.6 12.4 12.4 16.8 25.4

Name _____ Date _____ Class _____

LESSON 2-6 Practice A
Adding and Subtracting with Unlike Denominators

Name a common denominator for each sum or difference. Do not solve.

1. $\dfrac{1}{2} + \dfrac{3}{4}$
2. $\dfrac{1}{3} + \dfrac{4}{9}$
3. $\dfrac{2}{3} - \dfrac{3}{8}$
4. $\dfrac{1}{2} - \dfrac{1}{6}$

Add or subtract. Write answer in simplest form.

5. $\dfrac{1}{5} + \dfrac{1}{2}$
6. $\dfrac{3}{4} + \dfrac{5}{6}$
7. $\dfrac{7}{10} - \dfrac{1}{5}$
8. $\dfrac{3}{14} - \dfrac{4}{7}$

9. $3\dfrac{1}{3} - 1\dfrac{1}{6}$
10. $\dfrac{1}{4} + \dfrac{1}{2}$
11. $4\dfrac{1}{3} - 2\dfrac{7}{9}$
12. $2\dfrac{3}{5} + \left(-1\dfrac{7}{10}\right)$

13. $\dfrac{2}{5} - \dfrac{1}{2}$
14. $2\dfrac{1}{3} + 1\dfrac{1}{2}$
15. $3\dfrac{1}{4} + \left(-1\dfrac{5}{6}\right)$
16. $\dfrac{3}{4} - \dfrac{11}{12}$

Evaluate each expression for the given value of the variable.

17. $1\dfrac{7}{8} + x$ for $x = -2\dfrac{3}{4}$
18. $x - \dfrac{2}{3}$ for $x = \dfrac{1}{6}$
19. $x - \dfrac{1}{2}$ for $x = \dfrac{7}{8}$

20. $2\dfrac{2}{3} + x$ for $x = -1\dfrac{5}{9}$
21. $x - \dfrac{2}{5}$ for $x = \dfrac{9}{10}$
22. $x - \dfrac{6}{7}$ for $x = \dfrac{1}{2}$

23. Mr. Martanarie bought a new lamp and lamppost for his home. The pole was $6\dfrac{5}{8}$ ft tall and the lamp was $1\dfrac{1}{4}$ ft in height. How tall were the lamp and post together?

Name _____ Date _____ Class _____

LESSON 2-6 Practice B
Adding and Subtracting with Unlike Denominators

Add or subtract.

1. $\dfrac{2}{3} + \dfrac{1}{2}$

2. $\dfrac{3}{5} + \dfrac{1}{3}$

3. $\dfrac{3}{4} - \dfrac{1}{3}$

4. $\dfrac{1}{2} - \dfrac{5}{9}$

5. $\dfrac{5}{16} - \dfrac{5}{8}$

6. $\dfrac{7}{9} + \dfrac{5}{6}$

7. $\dfrac{7}{8} - \dfrac{1}{4}$

8. $\dfrac{5}{6} - \dfrac{3}{8}$

9. $2\dfrac{7}{8} + 3\dfrac{5}{12}$

10. $1\dfrac{2}{9} + 2\dfrac{1}{18}$

11. $3\dfrac{2}{3} - 1\dfrac{3}{5}$

12. $1\dfrac{5}{6} + (-2\dfrac{3}{4})$

13. $8\dfrac{1}{3} - 3\dfrac{5}{9}$

14. $5\dfrac{1}{3} + 1\dfrac{11}{12}$

15. $7\dfrac{1}{4} + (-2\dfrac{5}{12})$

16. $5\dfrac{2}{5} - 7\dfrac{3}{10}$

Evaluate each expression for the given value of the variable.

17. $2\dfrac{3}{8} + x$ for $x = 1\dfrac{5}{6}$

18. $x - \dfrac{2}{5}$ for $x = \dfrac{1}{3}$

19. $x - \dfrac{3}{10}$ for $x = \dfrac{3}{7}$

20. $1\dfrac{5}{8} + x$ for $x = -2\dfrac{1}{6}$

21. $x - \dfrac{3}{4}$ for $x = \dfrac{1}{6}$

22. $x - \dfrac{3}{10}$ for $x = \dfrac{1}{2}$

23. Ana worked $6\dfrac{1}{2}$ h on Monday, $5\dfrac{3}{4}$ h on Tuesday and $7\dfrac{1}{6}$ h on Friday. How many total hours did she work these three days?

Holt Mathematics

Name _____ Date _____ Class _____

LESSON 2-6 Practice C
Adding and Subtracting with Unlike Denominators

Add or subtract.

1. $\dfrac{7}{12} + \dfrac{5}{9}$

2. $\dfrac{7}{10} + \dfrac{4}{15}$

3. $\dfrac{7}{8} - \dfrac{11}{12}$

4. $\dfrac{15}{16} - \dfrac{5}{32}$

_____   _____   _____   _____

5. $1\dfrac{1}{18} + \left(-4\dfrac{15}{24}\right)$

6. $8\dfrac{3}{20} - 3\dfrac{7}{30}$

7. $8\dfrac{7}{10} + \left(-6\dfrac{2}{25}\right)$

8. $5\dfrac{11}{15} - 7\dfrac{5}{6}$

_____   _____   _____   _____

Evaluate each expression for the given value of the variable.

9. $10\dfrac{6}{25} + x$ for $x = -2\dfrac{3}{5}$

10. $x - \dfrac{5}{18}$ for $x = \dfrac{5}{6}$

_____   _____

11. $1\dfrac{5}{21} + x$ for $x = -5\dfrac{6}{7}$

12. $x - \dfrac{8}{11}$ for $x = \dfrac{13}{22}$

_____   _____

13. $14\dfrac{1}{15} + x$ for $x = -9\dfrac{3}{10}$

14. $x - \dfrac{7}{9}$ for $x = \dfrac{2}{27}$

_____   _____

15. A carpenter cuts a piece of wood that is $9\dfrac{5}{16}$ ft long into two pieces. One piece is $5\dfrac{3}{8}$ ft long. How long is the other piece?

_____

16. Before April 9, 2001, when the U.S. Securities and Exchange Commission ordered all U.S. stock markets to report stocks in decimals, the price of stock was reported in fractions. Under the fractional reporting system, what was the change in stock price if a stock opened the day at $31\dfrac{1}{8}$ and closed the day at $28\dfrac{5}{32}$?

_____

LESSON 2-6 Reteach
Adding and Subtracting with Unlike Denominators

To model $\frac{1}{2} + \frac{1}{3}$, use two rectangles of the same size and shape.

A. 1st rectangle: Shade $\frac{1}{2}$ vertically.

$\frac{1}{2}$

B. 2nd rectangle: Shade $\frac{1}{3}$ horizontally.

$\frac{1}{3}$

C. Separate the shaded portions into parts of equal size.

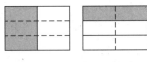

$\frac{1}{2} = \frac{3}{6}$ $\frac{1}{3} = \frac{2}{6}$

D. Use a new rectangle to show the sum.

$\frac{1}{2} + \frac{1}{3} = \frac{3}{6} + \frac{2}{6} = \frac{5}{6}$

Model $\frac{1}{2} + \frac{2}{5}$. Write the result.

1.

Model $\frac{1}{3} + \frac{3}{5}$. Write the result.

2.

Name _____ Date _____ Class _____

LESSON 2-6 Reteach
Adding and Subtracting with Unlike Denominators (continued)

To add fractions with different denominators, first write the fractions with common denominators. To find the LCD of denominators 5 and 6, list the multiples of each.

    Multiples of 5: 5, 10, 15, 20, 25, ㉚
    Multiples of 6: 12, 18, 24, ㉚
    So, the LCD of 5 and 6 is 30.

Complete to find the LCD for each set of denominators.

3. The LCD of 6 and 4 is: _____

   Multiples of 6: _____

   Multiples of 4: _____

4. The LCD of 3 and 7 is: _____

   Multiples of 3: _____

   Multiples of 7: _____

To add fractions with different denominators:

$$\text{Add: } \frac{1}{2} + \frac{1}{3} = \frac{1 \cdot 3}{2 \cdot 3} = \frac{3}{6}$$
$$\frac{1 \cdot 2}{3 \cdot 2} = \frac{2}{6}$$
$$= \frac{5}{6}$$

Complete to add fractions. Simplify.

5. $\frac{1}{4} = \frac{}{20}$
   $+ \frac{3}{5} = \frac{}{20}$
   = _____

6. $\frac{3}{4} = \frac{}{16}$
   $+ \frac{5}{16} = \frac{}{16}$
   = _____ = _____

7. $5\frac{1}{3} = 5\frac{}{24}$
   $+ 2\frac{5}{8} = 2\frac{}{24}$
   = _____

Add or subtract fractions. Simplify.

8. $\frac{1}{4} + \frac{7}{20} =$

9. $\frac{4}{9} - \frac{1}{5} =$

10. $\frac{8}{15} - \frac{1}{4} =$

Holt Mathematics

Name _____ Date _____ Class _____

LESSON 2-6 Challenge
Please Repeat That.

A decimal that repeats one digit is equivalent to a fraction with denominator 9.

$0.\overline{1} = \frac{1}{9}$ $0.\overline{2} = \frac{2}{9}$ $0.\overline{5} = \frac{5}{9}$

A decimal that repeats two digits is equivalent to a fraction with denominator 99.

$0.\overline{43} = \frac{43}{99}$ $0.\overline{61} = \frac{61}{99}$ $0.\overline{38} = \frac{38}{99}$

The pattern continues so that $0.\overline{681} = \frac{681}{999}$ and $0.\overline{24793} = \frac{24{,}793}{99{,}999}$.

Use a calculator to write each decimal equivalent.

1. $\frac{1}{90} =$ _____

2. $\frac{21}{990} =$ _____

3. $\frac{358}{9990} =$ _____

Predict the decimal equivalent of each fraction. Verify your results on a calculator.

4. $\frac{4}{90} =$ _____

5. $\frac{62}{990} =$ _____

6. $\frac{617}{9990} =$ _____

Write each fractional equivalent and simplify.

7. $0.\overline{7} =$

8. $0.\overline{08} =$

9. $0.00\overline{24} =$

_____   _____   _____

When one digit repeats but does not begin in the first decimal place, and the digit in the first place is other than 0, you must add fractions.

$0.3\overline{7} = 0.3 + 0.0\overline{7}$
$= \frac{3}{10} + \frac{7}{90}$
$= \frac{27}{90} + \frac{7}{90} = \frac{34}{90} = \frac{17}{45}$

Write each repeating decimal as the sum of two fractions. Find the sum and simplify. Verify.

10. $0.2\overline{8} =$

11. $0.25\overline{32} =$

_____   _____

12. $0.1\overline{27} =$

13. $0.75\overline{483} =$

_____   _____

49   Holt Mathematics

Problem Solving
LESSON 2-6 Adding and Subtracting with Unlike Denominators

Write the correct answer.

1. Nick Hysong of the United States won the Olympic gold medal in the pole vault in 2000 with a jump of 19 ft $4\frac{1}{4}$ inches, or $232\frac{1}{4}$ inches. In 1900, Irving Baxter of the United States won the pole vault with a jump of 10 ft $9\frac{7}{8}$ inches, or $129\frac{7}{8}$ inches. How much higher did Hysong vault than Baxter?

2. In the 2000 Summer Olympics, Ivan Pedroso of Cuba won the Long jump with a jump of 28 ft $\frac{3}{4}$ inches, or $336\frac{3}{4}$ inches. Alvin Kraenzlein of the United States won the long jump in 1900 with a jump of 23 ft $6\frac{7}{8}$ inches, or $282\frac{7}{8}$ inches. How much farther did Pedroso jump than Kraenzlein?

3. A recipe calls for $\frac{1}{8}$ cup of sugar and $\frac{3}{4}$ cup of brown sugar. How much total sugar is added to the recipe?

4. The average snowfall in Norfolk, VA for January is $2\frac{3}{5}$ inches, February $2\frac{9}{10}$ inches, March 1 inch, and December $\frac{9}{10}$ inches. If these are the only months it typically snows, what is the average snowfall per year?

Use the table at the right that shows the average snowfall per month in Vail, Colorado.

5. What is the average annual snowfall in Vail, Colorado?

   A $15\frac{13}{20}$ in.   C $187\frac{1}{10}$ in.

   B 153 in.   D $187\frac{4}{5}$ in.

6. The peak of the skiing season is from December through March. What is the average snowfall for this period?

   F $30\frac{19}{20}$ in.   H $123\frac{4}{5}$ in.

   G $123\frac{3}{5}$ in.   J 127 in.

Average Snowfall in Vail, CO

Month	Snowfall (in.)	Month	Snowfall (in.)
Jan	$36\frac{7}{10}$	July	0
Feb	$35\frac{7}{10}$	August	0
March	$25\frac{2}{5}$	Sept	1
April	$21\frac{1}{5}$	Oct	$7\frac{4}{5}$
May	4	Nov	$29\frac{7}{10}$
June	$\frac{3}{10}$	Dec	26

Name _____ Date _____ Class _____

LESSON 2-6 Reading Strategies
Use a Graphic Aid

It is easy to add and subtract fractions with **common denominators**.

3 eighths + 4 eighths = 7 eighths   8 ninths − 3 ninths = 5 ninths

$$\frac{3}{8} + \frac{4}{8} = \frac{7}{8} \qquad \frac{8}{9} - \frac{3}{9} = \frac{5}{9}$$

Adding fractions with unlike denominators requires more steps. The picture below will help you understand adding fractions with unlike denominators. $\frac{1}{2} + \frac{1}{4} = ?$

$$\frac{1}{2} + \frac{1}{4}$$

In order to add $\frac{1}{2} + \frac{1}{4}$, you must find a common denominator.

1. What are the denominators in this problem? _____

2. To find a common denominator, one-half can be changed into fourths. How many fourths are there in one-half? _____

3. Change $\frac{1}{2}$ to fourths. _____

4. You can now add, because you have a common denominator. _____

To subtract fractions with unlike denominators, you must find a common denominator. The picture below will help you understand finding a common denominator. $\frac{5}{6} - \frac{1}{3} = ?$

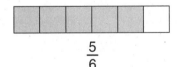
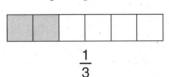

5. What are the denominators in this problem? _____

To find a common denominator, you will change to sixths.

6. How many sixths are in one-third? Write the fraction. _____

7. You can now subtract the fractions. _____

Name _____ Date _____ Class _____

LESSON 2-6 Puzzles, Twisters & Teasers
Just a Tad Difficult!

Add or subtract to solve each equation. Then use the answers to solve the riddle.

M $\dfrac{5}{12} + \dfrac{3}{7} = $ _____

T $\dfrac{1}{5} + \dfrac{7}{9} = $ _____

O $\dfrac{15}{16} - \dfrac{9}{10} = $ _____

H $\dfrac{1}{3} + 1\dfrac{1}{12} = $ _____

R $2\dfrac{1}{5} + 1\dfrac{8}{9} = $ _____

E $\dfrac{5}{8} + \dfrac{1}{6} = $ _____

K $\dfrac{5}{16} + \dfrac{2}{7} = $ _____

C $2\dfrac{1}{3} - 4\dfrac{7}{9} = $ _____

A $\dfrac{3}{4} - \dfrac{5}{16} = $ _____

N $1\dfrac{3}{4} + \left(-3\dfrac{13}{15}\right) = $ _____

Where do tadpoles change into frogs?

I ___ ___ ___ ___
$-2\dfrac{7}{60}$ $\dfrac{44}{45}$ $1\dfrac{5}{12}$ $\dfrac{19}{24}$

___ ___ ___ ___ ___   ___ ___ ___ ___
$-2\dfrac{4}{9}$ $4\dfrac{4}{45}$ $\dfrac{3}{80}$ $\dfrac{7}{16}$ $\dfrac{67}{112}$ $4\dfrac{4}{45}$ $\dfrac{3}{80}$ $\dfrac{3}{80}$ $\dfrac{71}{84}$

Practice A
2-7 Solving Equations with Rational Numbers

Solve.

1. $x + 1.2 = 4.6$

2. $a - 3.4 = 5$

3. $2.2m = 4.4$

4. $\dfrac{x}{1.3} = 2$

5. $6.7 + w = -1.1$

6. $\dfrac{n}{1.9} = -3.8$

7. $7.2 = -0.9y$

8. $k - 4.05 = 6.2$

9. $\dfrac{d}{-3.2} = -3.75$

10. $-\dfrac{2}{5} + x = \dfrac{2}{5}$

11. $\dfrac{1}{4}x = \dfrac{1}{2}$

12. $\dfrac{1}{3}a = \dfrac{3}{4}$

13. $x - \dfrac{3}{2} = \dfrac{1}{5}$

14. $x - \dfrac{3}{7} = -\dfrac{5}{7}$

15. $-\dfrac{5}{6}a = \dfrac{5}{8}$

16. Elisa can reach $77\dfrac{3}{4}$ in. high. The ceiling is $90\dfrac{1}{2}$ in. high. How much higher is the ceiling than Elisa's highest reach?

17. Nolan makes $10.60 an hour at his after-school job. Last week he worked 11.25 hr. How much was Nolan paid for the week?

Name _____ Date _____ Class _____

LESSON 2-7 Practice B
Solving Equations with Rational Numbers

Solve.

1. $x + 6.8 = 12.19$

2. $y - 10.24 = 5.3$

3. $0.05w = 6.25$

4. $\dfrac{a}{9.05} = 8.2$

5. $-12.41 + x = -0.06$

6. $\dfrac{d}{-8.4} = -10.2$

7. $-2.89 = 1.7m$

8. $n - 8.09 = -11.65$

9. $\dfrac{x}{5.4} = -7.18$

10. $\dfrac{7}{9} + x = 1\dfrac{1}{9}$

11. $\dfrac{6}{11}y = -\dfrac{18}{22}$

12. $\dfrac{7}{10}d = \dfrac{21}{20}$

13. $x - \left(-\dfrac{9}{14}\right) = \dfrac{5}{7}$

14. $x - \dfrac{15}{21} = 2\dfrac{6}{7}$

15. $-\dfrac{8}{15}a = \dfrac{9}{10}$

16. A recipe calls for $2\dfrac{1}{3}$ cups of flour and $1\dfrac{1}{4}$ cups of sugar. If the recipe is tripled, how much flour and sugar will be needed?

17. Daniel filled the gas tank in his car with 14.6 gal of gas. He then drove 284.7 mi before needing to fill up his tank with gas again. How many miles did the car get to a gallon of gasoline?

Name _____ Date _____ Class _____

LESSON 2-7 Practice C
Solving Equations with Rational Numbers

Solve.

1. $x + 102.8 = 89.06$

2. $62.5m = 2587.5$

3. $\dfrac{w}{38.7} = 51.06$

_____   _____   _____

4. $-10\dfrac{5}{18} + x = -12\dfrac{3}{10}$

5. $5\dfrac{2}{15}a = 3\dfrac{2}{25}$

6. $y - \left(-6\dfrac{1}{16}\right) = 11\dfrac{3}{40}$

_____   _____   _____

7. A photo that is $3\dfrac{1}{2}$ in. by $5\dfrac{1}{4}$ in. is enlarged to three times its original size. What are the new dimensions of the photo?

_____

8. Mars has two very small elliptical shaped moons Deimos and Phobus. They were discovered in August 1877, by Asaph Hall, an American Astronomer. The inner satellite is Phobos. It is 16.78 mi long. Deimos is the outer moon and is 9.32 mi long. What is the difference in the length of the two moons?

_____

9. Mr. Crowley bought lunch for himself and eight of his employees. Each had a sandwich platter that costs $5.85 and a drink that costs $1.25. Five of the employees had dessert that costs $1.50. Mr. Crowley gave the delivery person $80 and told her to keep the change as a tip. How much was the delivery person's tip?

_____

10. Two of the greatest rainfalls ever recorded were on July 4, 1956. In Unionville, Maryland it rained 1.23 in. in 1 min. In Curtea-de-Arges, Romania on July 7, 1889 it rained 8.1 in. in 20 min. If it had rained for 20 min in Unionville at its same record pace, what would be the difference between the two rainfall amounts?

_____

Reteach
2-7 Solving Equations with Rational Numbers

Solving equations with rational numbers is basically the same as solving equations with integers or whole numbers:
  Use inverse operations to isolate the variable.

$\frac{1}{4}z = -16$
$4 \cdot \frac{1}{4}z = -16 \cdot 4$ *Multiply each side by 4.*
$z = -64$

$y - \frac{3}{8} = \frac{7}{8}$
$+ \frac{3}{8} \quad + \frac{3}{8}$ *Add $\frac{3}{8}$ to each side.*
$y = \frac{10}{8} = 1\frac{2}{8} = 1\frac{1}{4}$

$x + 3.5 = -17.42$
$\quad - 3.5 \quad - 3.5$ *Subtract 3.5 from each side.*
$x = -20.92$

$-26t = 317.2$
$\frac{-26t}{-26} = \frac{317.2}{-26}$ *Divide each side by −26.*
$t = -12.2$

Tell what you would do to isolate the variable.

1. $x - 1.4 = 7.82$

2. $\frac{1}{4} + y = \frac{7}{4}$

3. $3z = 5$

_____   _____   _____

Solve each equation.

4. $14x = -129.5$

5. $\frac{1}{3}y = 27$

6. $265.2 = \frac{z}{22.1}$

_____   _____   _____

7. $x + 53.8 = -1.2$

8. $25 = \frac{1}{5}k$

9. $m - \frac{2}{3} = \frac{3}{5}$

_____   _____   _____

Name _____ Date _____ Class _____

LESSON 2-7 Challenge
Location, Location, Location

An equation that has a variable in the denominator of one or more of its terms is called a **fractional equation**.

One method of solution is to clear the equation of fractions by multiplying each side of the equation by the LCD.

$\frac{1}{2} + \frac{1}{x} = \frac{3}{5}$ The LCD of 2, x, and 5 is $10x$, with $x \neq 0$.

$10x\left(\frac{1}{2} + \frac{1}{x}\right) = 10x\left(\frac{3}{5}\right)$ Multiply each side by $10x$.

$10x \cdot \frac{1}{2} + 10x \cdot \frac{1}{x} = 10x \cdot \frac{3}{5}$ Distributive Property

$5x + 10 = 6x$ Simplify.

$5x - 5x + 10 = 6x - 5x$ Subtract $5x$ from each side.

$10 = x$

Check:

$\frac{1}{2} + \frac{1}{x} = \frac{3}{5}$

$\frac{1}{2} + \frac{1}{10} \stackrel{?}{=} \frac{3}{5}$ Substitute 10 for x in the original equation.

$\frac{5}{10} + \frac{1}{10} \stackrel{?}{=} \frac{3}{5}$ Do not repeat the method of solution.

$\frac{6}{10} \stackrel{?}{=} \frac{3}{5}$ Work each side separately.

$\frac{3}{5} = \frac{3}{5}$ ✓

Solve and check.

1. $\frac{4}{7} + \frac{2}{x} = \frac{2}{3}$

2. $\frac{10}{x} + \frac{8}{x} = 9$

3. $\frac{15}{x} = 7 + \frac{9}{2x}$

Name _____ Date _____ Class _____

LESSON 2-7 Problem Solving
Solving Equations with Rational Numbers

Write the correct answer.

1. In the last 150 years, the average height of people in industrialized nations has increased by $\frac{1}{3}$ foot. Today, American men have an average height of $5\frac{7}{12}$ feet. What was the average height of American men 150 years ago?

2. Jaime has a length of ribbon that is $23\frac{1}{2}$ in. long. If she plans to cut the ribbon into pieces that are $\frac{3}{4}$ in. long, into how many pieces can she cut the ribbon? (She cannot use partial pieces.)

3. Todd's restaurant bill for dinner was $15.55. After he left a tip, he spent a total of $18.00 on dinner. How much money did Todd leave for a tip?

4. The difference between the boiling point and melting point of Hydrogen is 6.47°C. The melting point of Hydrogen is −259.34°C. What is the boiling point of Hydrogen?

Choose the letter for the best answer.

5. Justin Gatlin won the Olympic gold in the 100-m dash in 2004 with a time of 9.85 seconds. His time was 0.95 seconds faster than Francis Jarvis who won the 100-m dash in 1900. What was Jarvis' time in 1900?

   A 8.95 seconds
   B 10.65 seconds
   C 10.80 seconds
   D 11.20 seconds

6. The balance in Susan's checking account was $245.35. After the bank deposited interest into the account, her balance went to $248.02. How much interest did the bank pay Susan?

   F $1.01
   G $2.67
   H $3.95
   J $493.37

7. After a morning shower, there was $\frac{17}{100}$ in. of rain in the rain gauge. It rained again an hour later and the rain gauge showed $\frac{1}{4}$ in. of rain. How much did it rain the second time?

   A $\frac{2}{25}$ in. C $\frac{21}{50}$ in.
   B $\frac{1}{6}$ in.   D $\frac{3}{8}$ in.

8. Two-third of John's savings account is being saved for his college education. If $2500 of his savings is for his college education, how much money in total is in his savings account?

   F $1666.67   H $4250.83
   G $3750   J $5000

Name _____ Date _____ Class _____

LESSON 2-7 Reading Strategies
Follow a Procedure

The rules for solving equations with rational numbers are the same as equations with whole numbers.

| Get the variable by itself. | → | Perform the same operation on both sides to keep the equation balanced. | → | Use the rules for computing rational numbers. |

Follow the steps above to help you solve $\frac{1}{4} + y = \frac{3}{4}$.

1. What is the first step to solve this equation?

   _____

2. What operation should you use?

   _____

3. Write an equation to show the subtraction of $\frac{1}{4}$ on both sides.

   _____

4. What is the value of y?

   _____

Follow the steps above to solve $x - 4.5 = 13$.

5. What is the first step to solve this equation?

   _____

6. What operation should you use?

   _____

7. Write an equation to show the addition of 4.5 to both sides.

   _____

8. Find the value of x.

   _____

Puzzles, Twisters & Teasers

LESSON 2-7 Math Book Issues!

Solve the equations. Then use the letters of the variables to answer the riddle.

1. $s - \frac{5}{9} = \frac{1}{9}$ $s = $ _____

2. $t + \frac{1}{3} = \frac{3}{4}$ $t = $ _____

3. $m - \frac{1}{12} = \frac{5}{12}$ $m = $ _____

4. $o + \frac{5}{9} = -\frac{1}{9}$ $o = $ _____

5. $e - 17.9 = 36.8$ $e = $ _____

6. $a - 2.1 = -4.5$ $a = $ _____

7. $l + \frac{4}{13} = \frac{12}{39}$ $l = $ _____

8. $0.04n = 0.252$ $n = $ _____

9. $b \div 3.2 = -6$ $b = $ _____

10. $\frac{5}{9} + y = \frac{6}{18}$ $y = $ _____

11. $r + \frac{5}{8} = -2\frac{3}{8}$ $r = $ _____

12. $p + 3.8 = -1.6$ $p = $ _____

Why did the student return her math book?

It had ___ ___ ___   ___ ___ ___ ___
       $\frac{5}{12}$ $-\frac{2}{3}$ $-\frac{2}{3}$ $\frac{1}{2}$ -2.4 6.3 $-\frac{2}{9}$

___ ___ ___ ___ ___ ___ ___ ___ .
-5.4 -3 $-\frac{2}{3}$ -19.2 0 54.7 $\frac{1}{2}$ $\frac{2}{3}$

Name _____ Date _____ Class _____

LESSON 2-8 Practice A
Solving Two-Step Equations

Describe the operation performed on both sides of the equation in steps 2 and 4.

1. $3x + 2 = 11$
   $3x + 2 - 2 = 11 - 2$ _____
   $3x = 9$
   $\frac{3x}{3} = \frac{9}{3}$ _____
   $x = 3$

2. $\frac{x}{4} - 1 = -2$
   $\frac{x}{4} - 1 + 1 = -2 + 1$ _____
   $\frac{x}{4} = -1$
   $4\left(\frac{x}{4}\right) = 4(-1)$ _____
   $x = -4$

Solve.

3. $2x + 3 = 9$

4. $\frac{x}{3} - 1 = 5$

5. $-3a + 4 = 7$

6. $\frac{x + 2}{2} = -3$

_____   _____   _____   _____

7. $5y - 2 = 28$

8. $2x - 7 = 7$

9. $\frac{w - 2}{5} = -1$

10. $2r + 1 = -1$

_____   _____   _____   _____

Write and solve a two-step equation to answer the question.

11. Pearson rented a moving van for 1 day. The total rental charge is $66.00. A daily rental costs $45.00 plus $0.25 per mile. How many miles did he drive the van?

_____

_____

_____

_____

_____

Name _____ Date _____ Class _____

LESSON 2-8 Practice B
Solving Two-Step Equations

Write and solve a two-step equation to answer the following questions.

1. The school purchased baseball equipment and uniforms for a total cost of $1762. The equipment costs $598 and the uniforms were $24.25 each. How many uniforms did the school purchase?

   _____
   _____
   _____
   _____

2. Carla runs 4 miles every day. She jogs from home to the school track, which is $\frac{3}{4}$ mile away. She then runs laps around the $\frac{1}{4}$-mile track. Carla then jogs home. How many laps does she run at the school?

   _____
   _____
   _____
   _____

Solve.

3. $\frac{a+5}{3} = 12$ 4. $\frac{x+2}{4} = -2$ 5. $\frac{y-4}{6} = -3$ 6. $\frac{k+1}{8} = 7$

7. $0.5x - 6 = -4$ 8. $\frac{x}{2} + 3 = -4$ 9. $\frac{1}{5}n + 3 = 6$ 10. $2a - 7 = -9$

11. $\frac{3x-1}{4} = 2$ 12. $-7.8 = 4.4 + 2r$ 13. $\frac{-4w+5}{-3} = -7$ 14. $1.3 - 5r = 7.4$

15. A phone call costs $0.58 for the first 3 minutes and $0.15 for each additional minute. If the total charge for the call was $4.78, how many minutes was the call? _____

16. Seventeen less than four times a number is twenty-seven. Find the number. _____

Copyright © by Holt, Rinehart and Winston.
All rights reserved.

62

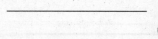

Name _____ Date _____ Class _____

LESSON 2-8 Practice C
Solving Two-Step Equations

Write an equation for each sentence, then solve it.

1. A number multiplied by five and increased by 3 is 28.

2. Eighteen decreased by 4 times a number is 62.

3. The sum of 3 times a number and 7, divided by 5, is 17.

4. The quotient of a number and 5, minus 2, is 8.

Solve.

5. $-13 = -3x + 14$

6. $2w - 5 = 4$

7. $\dfrac{x-5}{3} = -10$

8. $\dfrac{2}{3}n - 7 = 19$

9. $1.4x + 0.8 = -1.3$

10. $\dfrac{1}{4}y + \dfrac{3}{4} = 2$

11. $15n - 62 = -17$

12. $\dfrac{3}{8}a - 4 = -\dfrac{1}{4}$

13. $\dfrac{2d+9}{6} = 11$

14. $24.5 = 16.1 - 2.4r$

15. $\dfrac{7x-2}{6} = -5$

16. $\dfrac{5}{6} - \dfrac{a}{4} = \dfrac{1}{3}$

17. Larissa is planning for a trip that cost $2145. She has $952.50 saved and is going to set aside $\dfrac{1}{2}$ of her weekly salary from her part-time job. Larissa earns $265 a week. How many weeks will it take her to earn the rest of the money needed for the trip? _____

18. Noel bought a printer for $10 less than half its original price. If Noel paid $88 for the printer, what was the original price? _____

Name _____ Date _____ Class _____

LESSON 2-8 Reteach
Solving Two-Step Equations

To solve an equation, it is important to first note how it is formed.
Then, work backward to undo each operation.

$4z + 3 = 15$	$\dfrac{z}{4} - 3 = 7$	$\dfrac{z + 3}{4} = 7$
The variable is multiplied by 4 and then 3 is added.	The variable is divided by 4 and then 3 is subtracted.	3 is added to the variable and then the result is divided by 4.
To solve, first subtract 3 and then divide by 4.	To solve, first add 3 and then multiply by 4.	To solve, multiply by 4 and then subtract 3.

Describe how each equation is formed.
Then, tell the steps needed to solve.

1. $3x - 5 = 7$

   The variable is _____ and then _____.

   To solve, first _____ and then _____.

2. $\dfrac{x}{3} + 5 = 7$

   The variable is _____ and then _____.

   To solve, first _____ and then _____.

3. $\dfrac{x + 5}{3} = 7$

   5 is _____ and then the result is _____.

   To solve, first _____ and then _____.

4. $10 = -3x - 2$

   The variable is _____ and then _____.

   To solve, first _____ and then _____.

5. $10 = \dfrac{x - 2}{5}$

   2 is _____ the variable and then the result is _____.

   To solve, first _____ and then _____.

Reteach
LESSON 2-8 Solving Two-Step Equations (continued)

To isolate the variable, work backward using inverse operations.

The variable is multiplied by 2 and then 3 is added.		The variable is divided by 2 and then 3 is subtracted.	
$2x + 3 = 11$	To undo addition,	$\frac{x}{2} - 3 = 11$	To undo subtraction,
$\underline{-3 \quad -3}$	subtract 3.	$\underline{+3 \ +3}$	add 3.
$2x \quad = 8$	To undo multiplication,	$\frac{x}{2} = 14$	To undo division,
$\frac{2x}{2} = \frac{8}{2}$	divide by 2.	$2 \cdot \frac{x}{2} = 2 \cdot 14$	multiply by 2.
$x = 4$		$x = 28$	

Check: Substitute 4 for x.
$2(4) + 3 \stackrel{?}{=} 11$
$8 + 3 \stackrel{?}{=} 11$
$11 = 11$ ✓

Check: Substitute 28 for x.
$\frac{28}{2} - 3 \stackrel{?}{=} 11$
$14 - 3 \stackrel{?}{=} 11$
$11 = 11$ ✓

Complete to solve and check each equation.

6. $3t + 7 = 19$ To undo addition, subtract. **Check:** $3t + 7 = 19$
   ___ ___ $3(\_\_\_) + 7 \stackrel{?}{=} 19$ Substitute for t.
   $3t = \_\_\_$ To undo multiplication, divide. ___ $+ 7 \stackrel{?}{=} 19$
   $3t \div \_\_\_ = \_\_\_ \div \_\_\_$ _____
   $t = \_\_\_$

7. $\frac{w}{3} - 7 = 5$ To undo subtraction, add. **Check:** $\frac{w}{3} - 7 = 5$
   ___ ___   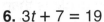 $- 7 \stackrel{?}{=} 5$ Substitute.
   $\frac{w}{3} = \_\_\_$ To undo division, multiply. ___ $- 7 \stackrel{?}{=} 5$
   $\_\_\_ \cdot \frac{w}{3} = \_\_\_ \cdot 12$ _____
   $w = \_\_\_$

8. $\frac{z-3}{2} = 8$ To undo division, multiply. **Check:** $\frac{z-3}{2} = 8$
   $\_\_\_ \cdot \frac{z-3}{2} = \_\_\_ \cdot 8$ $\frac{\_\_\_ - 3}{2} \stackrel{?}{=} 8$ Substitute.
   $z - 3 = \_\_\_$ To undo subtraction, add. $\frac{\_\_\_}{2} \stackrel{?}{=} 8$
   $z \_\_\_ = \_\_\_$ _____

LESSON 2-8 Challenge
Work It Algebraically!

An equation may be used to solve a problem involving probability.

A bag contains marbles of four colors: red, white, blue, and yellow. There are 3 more blue marbles than red, and 48 marbles in all. How many blue marbles are there if, in one draw, the probability of getting a blue marble is $\frac{5}{12}$?

Let x = the number of blue marbles.

$P(\text{blue}) = \dfrac{\text{number of successes}}{\text{total number}}$

$\dfrac{5}{12} = \dfrac{x}{48}$

$\dfrac{5}{12} \cdot 48 = 48 \cdot \dfrac{x}{48}$ Multiply by 48.

$20 = x$

So, there are 20 blue marbles in the bag.

Write and solve an equation for each problem.

1. A box has four kinds of candies: lime, orange, cherry, and mint. There are 9 more lime candies than orange, and 36 candies in all. How many lime candies are there if, in one draw, the probability of getting a lime candy is $\frac{4}{9}$?

2. A carton has four kinds of cookies: lemon, mint, vanilla, and chocolate. There are 7 fewer mint cookies than lemon, and 64 cookies in all. How many mint cookies are there if, in one draw, the probability of getting a lemon cookie is $\frac{5}{8}$?

There are _____ lime candies.

There are _____ mint cookies.

Problem Solving
2-8 Solving Two-Step Equations

The chart below describes three different long distance calling plans. Jamie has budgeted $20 per month for long distance calls. Write the correct answer.

1. How many minutes will Jamie be able to use per month with plan A? Round to the nearest minute.

   _____

Plan	Monthly Access Fee	Charge per minute
A	$3.95	$0.08
B	$8.95	$0.06
C	$0	$0.10

2. How many minutes will Jamie be able to use per month with plan B? Round to the nearest minute.

   _____

3. How many minutes will Jamie be able to use per month with plan C? Round to the nearest minute.

   _____

4. Which plan is the best deal for Jamie's budget?

   _____

5. Nolan has budgeted $50 per month for long distance. Which plan is the best deal for Nolan's budget?

   _____

The table describes four different car loans that Susana can get to finance her new car. The total column gives the amount she will end up paying for the car including the down payment and the payments with interest. Choose the letter for the best answer.

6. How much will Susana pay each month with loan A?
   A $252.04   C $330.35
   B $297.02   D $353.68

Loan	Down Payment	Number of Months	Total
A	$2000	60	$19,821.20
B	$1000	48	$19,390.72
C	$0	60	$20,197.20

7. How much will Susana pay each month with loan B?
   F $300.85   H $323.17
   G $306.50   J $383.14

8. How much will Susana pay each month with loan C?
   A $336.62   C $369.95
   B $352.28   D $420.78

9. Which loan will give Susana the smallest monthly payment?
   F Loan A   H Loan C
   G Loan B   J They are equal

Name _____ Date _____ Class _____

Reading Strategies
LESSON 2-8 *Analyze Information*

Break a problem into parts and analyze the information.

> Jill has $8 in her pocket now. She had $20 when she left for the movies. How much money did she spend?

Answer the questions in Exercises 1–4 to solve this problem.

1. How much money did Jill start with?

   _____

2. How much money does Jill have left?

   _____

3. What is the difference between these two amounts?

   _____

4. How much money did Jill spend?

   _____

> Mark paid $45 at the music store for 3 CDs and a pack of batteries, before tax. The batteries cost $6. How much did Mark pay for each of the CDs?

Answer the questions in Exercises 5–9 to solve this problem.

5. How much did Mark spend at the music store?

   _____

6. How much did Mark spend on batteries?

   _____

7. What is the difference between these two amounts?

   _____

8. Since Mark paid $39 for CDs, divide $39 by 3.

   _____

9. How much did Mark pay for each CD?

   _____

Puzzles, Twisters & Teasers
LESSON 2-8
Have a Ball!

Solve the equations. Match the letters of the variables to the answers to solve the riddle.

1. $7 + \frac{l}{5} = -4$ $l =$ _____

2. $46 - 3t = -23$ $t =$ _____

3. $8 = 6 + \frac{a}{4}$ $a =$ _____

4. $6h + 24 = 0$ $h =$ _____

5. $9 = -5b - 23$ $b =$ _____

6. $15 - 3e = -6$ $e =$ _____

7. $15w - 4 = 41$ $w =$ _____

8. $6s + 3 = -27$ $s =$ _____

9. $14o - 17 = 39$ $o =$ _____

10. $\frac{n}{-3} - 2 = 8$ $n =$ _____

Where do penguins go to dance?

A ___   T ___ ___   ___ ___ ___ ___ ___   L ___
23   -4   7   -5 -30 4 3 -6.4   8   -55

Name _____ Date _____ Class _____

CHAPTER 2 — Teacher Tool
Cutouts

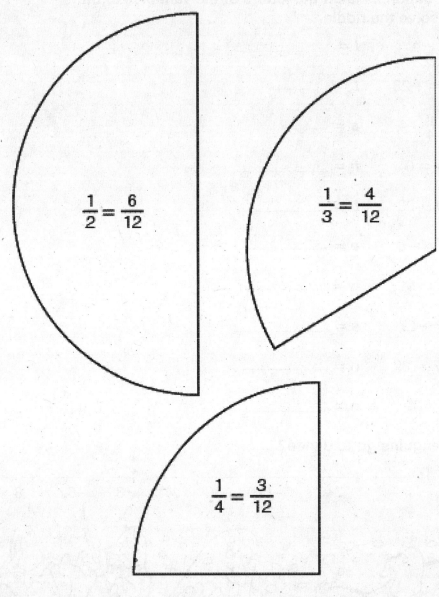
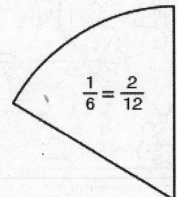
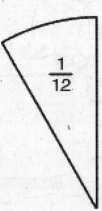

Copyright © by Holt, Rinehart and Winston.
All rights reserved.

Holt Mathematics

LESSON 2-1 Practice A
Rational Numbers

Simplify.

1. $\frac{4}{12}$ = $\frac{1}{3}$
2. $\frac{5}{15}$ = $\frac{1}{3}$
3. $-\frac{2}{8}$ = $-\frac{1}{4}$
4. $\frac{6}{24}$ = $\frac{1}{4}$
5. $\frac{14}{24}$ = $\frac{7}{12}$
6. $-\frac{15}{35}$ = $-\frac{3}{7}$
7. $\frac{10}{21}$ = $\frac{10}{21}$
8. $-\frac{16}{36}$ = $-\frac{4}{9}$

Write each decimal as a fraction in simplest form.

9. 0.4 = $\frac{2}{5}$
10. −0.35 = $-\frac{7}{20}$
11. 0.105 = $\frac{21}{200}$
12. 1.2 = $1\frac{1}{5}$
13. −0.85 = $-\frac{17}{20}$
14. 0.325 = $\frac{13}{40}$
15. 0.002 = $\frac{1}{500}$
16. 2.3 = $2\frac{3}{10}$
17. 0.28 = $\frac{7}{25}$
18. −1.25 = $-1\frac{1}{4}$
19. 0.064 = $\frac{8}{125}$
20. 0.0075 = $\frac{3}{400}$

Write each fraction as a decimal.

21. $\frac{1}{9}$ = $0.\overline{111}$
22. $\frac{9}{16}$ = 0.5625
23. $-\frac{11}{20}$ = −0.55
24. $\frac{6}{5}$ = 1.2
25. $\frac{2}{15}$ = $0.1\overline{33}$
26. $-2\frac{7}{12}$ = $-2.58\overline{33}$
27. $\frac{3}{100}$ = 0.03
28. $5\frac{8}{25}$ = 5.32

29. Make up a fraction that cannot be simplified that has 12 as its denominator.
sample answer: $\frac{5}{12}$, $\frac{1}{12}$, $\frac{7}{12}$, $\frac{11}{12}$

LESSON 2-1 Practice B
Rational Numbers

Simplify.

1. $\frac{6}{9}$ = $\frac{2}{3}$
2. $\frac{48}{96}$ = $\frac{1}{2}$
3. $\frac{13}{52}$ = $\frac{1}{4}$
4. $-\frac{7}{28}$ = $-\frac{1}{4}$
5. $\frac{15}{40}$ = $\frac{3}{8}$
6. $-\frac{4}{48}$ = $-\frac{1}{12}$
7. $-\frac{14}{63}$ = $-\frac{2}{9}$
8. $\frac{12}{72}$ = $\frac{1}{6}$

Write each decimal as a fraction in simplest form.

9. 0.72 = $\frac{18}{25}$
10. 0.058 = $\frac{29}{500}$
11. −1.65 = $-1\frac{13}{20}$
12. 2.1 = $2\frac{1}{10}$
13. 0.036 = $\frac{9}{250}$
14. −4.06 = $-4\frac{3}{50}$
15. 2.305 = $2\frac{61}{200}$
16. 0.0064 = $\frac{4}{625}$
17. −0.60 = $-\frac{3}{5}$
18. 6.95 = $6\frac{19}{20}$
19. 0.016 = $\frac{2}{125}$
20. 0.0005 = $\frac{1}{2000}$

Write each fraction as a decimal.

21. $\frac{1}{8}$ = 0.125
22. $\frac{8}{3}$ = $2.6\overline{66}$
23. $\frac{14}{15}$ = $0.9\overline{33}$
24. $\frac{16}{5}$ = 3.2
25. $\frac{11}{16}$ = 0.6875
26. $\frac{7}{9}$ = $0.\overline{777}$
27. $\frac{4}{5}$ = 0.8
28. $\frac{31}{25}$ = 1.24

29. Make up a fraction that cannot be simplified that has 24 as its denominator.
sample answer: $\frac{5}{24}$

LESSON 2-1 Practice C
Rational Numbers

Write each decimal as a fraction in simplest form.

1. 0.9 = $\frac{9}{10}$
2. 2.5 = $2\frac{1}{2}$
3. −0.36 = $-\frac{9}{25}$
4. −0.215 = $-\frac{43}{200}$
5. −4.02 = $-4\frac{1}{50}$
6. 0.0085 = $\frac{17}{2000}$
7. 1.006 = $1\frac{3}{500}$
8. 0.45 = $\frac{9}{20}$

Write each fraction as a decimal.

9. $\frac{9}{15}$ = 0.6
10. $-\frac{22}{50}$ = −0.44
11. $\frac{45}{16}$ = 2.8125
12. $-\frac{18}{90}$ = −0.2
13. $\frac{15}{80}$ = 0.1875
14. $\frac{21}{126}$ = $0.1\overline{66}$
15. $\frac{19}{12}$ = $1.58\overline{33}$
16. $\frac{39}{20}$ = 1.95

17. Make up a fraction that cannot be simplified that has 48 as its denominator.
sample answer: $\frac{5}{48}$

18. a. Simplify each fraction below.
   b. Write the denominator of each simplified fraction as the product of prime factors.
   c. Write each fraction as a decimal. Label each as a terminating or repeating decimal.

   $\frac{4}{36}$ $\frac{5}{40}$ $\frac{10}{25}$

   a. $\frac{1}{9}$ a. $\frac{1}{8}$ a. $\frac{2}{5}$
   b. $\frac{1}{3 \cdot 3}$ b. $\frac{1}{2 \cdot 2 \cdot 2}$ b. $\frac{2}{5}$
   c. $0.\overline{111}$; repeating   c. 0.125; terminating   c. 0.4; terminating

---

LESSON 2-1 Reteach
Rational Numbers

A **rational number** is a *ratio* of two integers.

Rational Number = $\frac{\text{Integer}}{\text{Integer}}$ ← Numerator / ← Denominator

The set of rational numbers contains:
- all integers
- all fractions
- decimals that repeat, such as $0.4\overline{6}$
- decimals that terminate, such as 3.5

Rational Numbers: Integers, Fractions, Repeating Decimals, Terminating Decimals

To simplify a fraction, divide numerator and denominator by the highest common factor.
$\frac{5}{15} = \frac{5 \div 5}{15 \div 5} = \frac{1}{3}$

Complete to simplify each fraction.

1. $\frac{8}{16} = \frac{8 \div 8}{16 \div 8} = \frac{1}{2}$
2. $\frac{15}{45} = \frac{15 \div 15}{45 \div 15} = \frac{1}{3}$
3. $\frac{12}{30} = \frac{12 \div 6}{30 \div 6} = \frac{2}{5}$
4. $\frac{12}{24} = \frac{12 \div 12}{24 \div 12} = \frac{1}{2}$
5. $\frac{5}{35} = \frac{5 \div 5}{35 \div 5} = \frac{1}{7}$
6. $\frac{14}{49} = \frac{14 \div 7}{49 \div 7} = \frac{2}{7}$

Simplify each fraction.

7. $\frac{8}{56} = \frac{1}{7}$
8. $\frac{15}{50} = \frac{3}{10}$
9. $\frac{8}{36} = \frac{2}{9}$

To write a decimal as a fraction, use the number of decimal places to get the denominator. Then simplify.
$0.4 = \frac{4}{10} = \frac{4 \div 2}{10 \div 2} = \frac{2}{5}$

Complete to write each decimal as a fraction in simplest form.

10. $0.25 = \frac{25}{100} = \frac{25 \div 25}{100 \div 25} = \frac{1}{4}$
11. $0.375 = \frac{375}{1000} = \frac{375 \div 125}{1000 \div 125} = \frac{3}{8}$

Write each decimal as a fraction in simplest form.

12. $0.55 = \frac{11}{20}$
13. $0.32 = \frac{8}{25}$

Holt Mathematics

LESSON 2-1 Reteach
Rational Numbers (continued)

To write a fraction as a decimal, divide numerator by denominator.

A decimal may terminate.

$\frac{3}{4} = 4\overline{)3.00}$ → 0.75
−28
 20
−20
  0

A decimal may repeat.

$\frac{1}{3} = 3\overline{)1.00}$ → 0.3
−9
 10
 −9
  1

Complete to write each fraction as a decimal.

14. $\frac{15}{4} = 4\overline{)15.00}$ → **3.75**
15. $\frac{5}{6} = 6\overline{)5.00}$ → **0.8̄3**
16. $\frac{11}{3} = 3\overline{)11.00}$ → **3.6̄6̄**

Write each fraction as a decimal.

17. $\frac{5}{2} =$ **2.5**
18. $\frac{15}{8} =$ **1.875**
19. $\frac{28}{6} =$ **4.6̄**
20. $\frac{22}{4} =$ **5.5**
21. $\frac{62}{12} =$ **5.16̄**
22. $\frac{105}{10} =$ **10.5**

LESSON 2-1 Challenge
Encore, Encore, ...

Explore some patterns with repeating decimals. Use a calculator to write each decimal equivalent.

1. $\frac{1}{9} =$ **0.1̄**   2. $\frac{2}{9} =$ **0.2̄**   3. $\frac{3}{9} =$ **0.3̄**

Predict the decimal equivalent of each fraction. Verify your results on a calculator.

4. $\frac{4}{9} =$ **0.4̄**   5. $\frac{6}{9} =$ **0.6̄**   6. $\frac{8}{9} =$ **0.8̄**

Write each fractional equivalent.

7. $0.\bar{5} =$ **$\frac{5}{9}$**   8. $0.\bar{7} =$ **$\frac{7}{9}$**   9. $0.\bar{9} =$ **$\frac{9}{9} = 1$**

Use a calculator to write each decimal equivalent.

10. $\frac{42}{99} =$ **0.4̄2̄**   11. $\frac{358}{999} =$ **0.358**   12. $\frac{4276}{9999} =$ **0.4276**

Predict the decimal equivalent of each fraction.

13. $\frac{76}{99} =$ **0.7̄6̄**   14. $\frac{732}{999} =$ **0.732**   15. $\frac{1957}{9999} =$ **0.1957**

Write each fractional equivalent.

16. $0.\overline{45} =$ **$\frac{45}{99}$**   17. $0.\overline{148} =$ **$\frac{148}{999}$**   18. $0.\overline{7213} =$ **$\frac{7213}{9999}$**

19. Summarize your observations.

A single digit repeating decimal equates to a fraction with denominator 9, a 2-digit repeating decimal to denominator 99, and so on.

LESSON 2-1 Problem Solving
Rational Numbers

Write the correct answer.

1. Fill in the table below which shows the sizes of drill bits in a set.

2. Do the drill bit sizes convert to repeating or terminating decimals?
Terminating decimals

13-Piece Drill Bit Set

Fraction	Decimal	Fraction	Decimal	Fraction	Decimal
$\frac{1}{4}$"	0.25	$\frac{11}{64}$"	0.171875	$\frac{3}{32}$"	0.09375
$\frac{15}{64}$"	0.234375	$\frac{5}{32}$"	0.15625	$\frac{5}{64}$"	0.078125
$\frac{7}{32}$"	0.21875	$\frac{9}{64}$"	0.140625	$\frac{1}{16}$"	0.0625
$\frac{13}{64}$"	0.203125	$\frac{1}{8}$"	0.125		
$\frac{3}{16}$"	0.1875	$\frac{7}{64}$"	0.109375		

Use the table at the right that lists the world's smallest nations. Choose the letter for the best answer.

3. What is the area of Vatican City expressed as a fraction in simplest form?
A $\frac{8}{50}$ C $\frac{17}{1000}$
B $\frac{4}{25}$ **D $\frac{17}{100}$**

World's Smallest Nations

Nation	Area (square miles)
Vatican City	0.17
Monaco	0.75
Nauru	8.2

4. What is the area of Monaco expressed as a fraction in simplest form?
F $\frac{75}{100}$ **H $\frac{3}{4}$**
G $\frac{15}{20}$ J $\frac{2}{3}$

5. What is the area of Nauru expressed as a mixed number?
A $8\frac{1}{50}$ C $8\frac{2}{5}$
B $8\frac{2}{50}$ **D $8\frac{1}{5}$**

6. The average annual precipitation in Miami, FL is 57.55 inches. Express 57.55 as a mixed number.
F $57\frac{11}{20}$ H $57\frac{5}{10}$
G $57\frac{55}{1000}$ J $57\frac{1}{20}$

7. The average annual precipitation in Norfolk, VA is 45.22 inches. Express 45.22 as a mixed number.
A $45\frac{11}{50}$ C $45\frac{11}{20}$
B $45\frac{22}{1000}$ D $45\frac{1}{5}$

LESSON 2-1 Reading Strategies
Use a Graphic Organizer

Definition
The set of numbers that can be written in the form $\frac{a}{b}$, where a and b are integers and b does not equal 0.

Facts
Fractions are rational numbers.
Decimals that terminate or repeat are rational numbers.
Whole numbers are rational numbers.
Integers are rational numbers.
0 is a rational number.

Examples
$2 = \frac{2}{1}$
$\frac{7}{8}$
$0.37 = \frac{37}{100}$
$4\frac{1}{4} = \frac{17}{4}$
$-5 = -\frac{5}{1}$

Rational Numbers

Non-examples
$\sqrt{3}$ (the square root of 3)
π (3.14159...)
$\sqrt{3}$ and π cannot be written as decimals that terminate or repeat.

Use the chart to answer the following questions.

1. What is a rational number?
any number that can be written in the form $\frac{a}{b}$

2. Is 0.62 a rational number? Why or why not?
yes, because it can be written as $\frac{62}{100}$ — decimal that terminates

3. Is $2\frac{1}{3}$ a rational number? Why or why not?
yes, because it can be written as $\frac{7}{3}$

4. Is $\sqrt{7}$ a rational number? Why or why not?
no, because it cannot be written as a decimal that terminates or repeats.

5. Is −8 a rational number? Why or why not?
yes, because it can be written as $\frac{-8}{1}$ — integers

6. Is 0 a rational number? Why or why not?
yes, because it can be written $\frac{0}{1}$ — 0 is a rational #

LESSON 2-1 Puzzles, Twisters & Teasers
Let's Be Rational!

Circle words from the list in the word search. Then find a word that answers the riddle. Circle it and write it on the line.

rational equivalent numerator denominator relatively
prime simplify integer nonzero factor

```
H O L E Q U I V A L E N T N C
M I R U J I N O N Z E R O U Y
O P R F A C T O R X D E B M F
Q W A U I O E W E R T Y L E I
O T T L P J G O K I Y T R R L
B Y I M J U E W E R T Y U A P
N J O A S P R I M E T Y U T M
D E N O M I N A T O R I L O I
X W A R E L A T I V E L Y R S
P L L Q A Z X S W E D C V F R
```

What can you put in a bucket of water to make it lighter?

a H   O   L   E

LESSON 2-2 Practice A
Comparing and Ordering Rational Numbers

Compare. Write <, >, or =.

1. $\frac{3}{4}$ [>] $\frac{5}{7}$
   $\frac{21}{28}$ [>] $\frac{20}{28}$
   ↓   ↓
   $\frac{3}{4}$ [>] $\frac{5}{7}$

2. $-\frac{2}{5}$ [<] $-\frac{3}{8}$
   $-\frac{16}{40}$ [<] $-\frac{15}{40}$, so
   ↓   ↓
   $-\frac{2}{5}$ [<] $-\frac{3}{8}$

3. 0.3 [>] $\frac{1}{4}$
   0.3 [>] 0.25
   ↓   ↓
   0.3 [>] $\frac{1}{4}$

4. $\frac{1}{6}$ [<] $\frac{1}{3}$
   ↓   ↓
   $\frac{1}{6}$ [<] $\frac{2}{6}$
   ↓   ↓
   $\frac{1}{6}$ [<] $\frac{1}{3}$

5. 0.09 [<] $\frac{1}{2}$
   ↓   ↓
   0.09 [<] 0.5
   ↓   ↓
   0.09 [<] $\frac{1}{2}$

6. $-\frac{3}{5}$ [=] -0.6
   ↓   ↓
   -0.6 [=] -0.6
   ↓   ↓
   $-\frac{3}{5}$ [=] -0.6

7. $\frac{4}{5}$ [>] $\frac{7}{10}$
8. $-\frac{1}{4}$ [>] $-\frac{3}{4}$
9. $-\frac{2}{3}$ [<] $-\frac{1}{4}$
10. $\frac{5}{8}$ [<] $\frac{5}{6}$
11. $\frac{7}{9}$ [>] $\frac{2}{3}$
12. $-\frac{3}{5}$ [<] $-\frac{1}{10}$
13. $\frac{1}{5}$ [>] $\frac{1}{8}$
14. $-1\frac{4}{5}$ [<] -1.3
15. $1\frac{2}{3}$ [>] $1\frac{2}{5}$

16. Trail A is $2\frac{2}{5}$ miles long. Trail B is $\frac{1}{4}$ mile long. Trail C is $1\frac{9}{10}$ miles long. Trail D is 2.05 miles long. List the lengths of the trails from shortest to longest.
   $\frac{1}{4}$ mile, $1\frac{9}{10}$ miles, 2.05 miles, $2\frac{2}{5}$ miles

---

LESSON 2-2 Practice B
Comparing and Ordering Rational Numbers

Compare. Write <, >, or =.

1. $\frac{1}{8}$ [>] $\frac{1}{10}$
2. $\frac{3}{5}$ [<] $\frac{7}{10}$
3. $-\frac{1}{3}$ [>] $-\frac{3}{4}$
4. $\frac{5}{6}$ [>] $\frac{3}{4}$
5. $-\frac{2}{7}$ [>] $-\frac{1}{2}$
6. $1\frac{2}{9}$ [<] $1\frac{2}{3}$
7. $-\frac{8}{9}$ [<] $-\frac{3}{10}$
8. $-\frac{4}{5}$ [=] $-\frac{8}{10}$
9. 0.08 [<] $\frac{3}{10}$
10. $\frac{11}{15}$ [=] $0.7\overline{3}$
11. $2\frac{4}{9}$ [<] $2\frac{3}{4}$
12. $-\frac{5}{8}$ [<] -0.58
13. $3\frac{1}{4}$ [<] 3.3
14. $-\frac{1}{6}$ [>] $-\frac{1}{9}$
15. 0.75 [=] $\frac{3}{4}$
16. $-2\frac{1}{8}$ [<] -2.1
17. $1\frac{1}{2}$ [>] 1.456
18. $-\frac{3}{5}$ [=] -0.6

19. On Monday, Gina ran 1 mile in 9.3 minutes. Her times for running 1 mile on each of the next four days, relative to her time on Monday, were $-1\frac{2}{3}$ minutes, -1.45 minutes, -1.8 minutes, and $-1\frac{3}{8}$ minutes. List these relative times in order from least to greatest.
   -1.8 minutes, $-1\frac{2}{3}$ minutes, -1.45 minutes, $-1\frac{3}{8}$ minutes

20. Trail A is 3.1 miles long. Trail C is $3\frac{1}{4}$ miles long. Trail B is longer than Trail A but shorter than Trail C. What is a reasonable distance for the length of Trail B?
   Possible answer: 3.2 miles

LESSON 2-2 Practice C
Comparing and Ordering Rational Numbers

Compare. Write <, >, or =.

1. $\frac{1}{9}$ [>] $\frac{1}{20}$
2. $\frac{3}{8}$ [<] $\frac{5}{6}$
3. $-\frac{1}{6}$ [>] $-\frac{2}{3}$
4. $\frac{11}{20}$ [=] 0.55
5. $-\frac{4}{9}$ [>] $-\frac{1}{2}$
6. $1\frac{3}{5}$ [>] 1.35
7. $-\frac{5}{9}$ [<] -0.45
8. $-1\frac{7}{8}$ [<] -1.875
9. $\frac{5}{4}$ [<] 1.4
10. $-1\frac{1}{5}$ [<] -1.06
11. 4.00 [<] $\frac{24}{4}$
12. $-\frac{5}{12}$ [>] -0.56

Write a fraction or decimal that has a value between the given numbers. Possible answers are given for Ex. 13–18.

13. $\frac{1}{6}$ and $\frac{1}{5}$
   0.18
14. 0.7 and 0.71
   0.701
15. $-\frac{1}{6}$ and 0.2
   0

16. $-\frac{1}{2}$ and $-\frac{1}{4}$
   $-\frac{1}{3}$
17. 1.45 and 1.46
   1.455
18. $\frac{2}{3}$ and 0.75
   0.7

19. The students in one English class are reading the same book. Last night, James read $\frac{1}{4}$ of the book. Jennie read $\frac{3}{8}$ of the book. Kyle read 0.4 of the book, and Talia read 0.33 of the book. List the numbers in order from least to greatest. Who read the greatest number of pages last night?
   $\frac{1}{4}$, 0.33, $\frac{3}{8}$, 0.4; Kyle

20. Melanie ran $2\frac{5}{8}$ miles on Monday. On Friday she ran 2.8 miles. On Wednesday, she ran further than on Monday but not as far as on Friday. What is a reasonable fraction length for the distance Melanie ran on Wednesday? What is a reasonable decimal length for the distance?
   Possible answer: $2\frac{3}{4}$ miles; 2.7 miles

Holt Mathematics

LESSON 2-2 Reteach
Comparing and Ordering Rational Numbers

You can use number lines to compare two fractions that have different denominators.

Compare $\frac{3}{8}$ and $\frac{2}{3}$.   Compare $-\frac{3}{4}$ and $-\frac{5}{6}$.

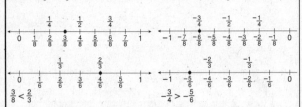

$\frac{3}{8} < \frac{2}{3}$ $-\frac{3}{4} > -\frac{5}{6}$

Use the number lines above. Write < or >.

1. $\frac{1}{4}$ > $\frac{1}{6}$ 2. $\frac{5}{6}$ > $\frac{1}{2}$ 3. $-\frac{3}{2}$ < $-\frac{1}{4}$ 4. $-\frac{5}{6}$ > $-\frac{5}{8}$

You can also use number lines to compare a fraction and a decimal.

Compare 0.2 and $\frac{1}{3}$.   Compare $-\frac{5}{8}$ and -0.9.

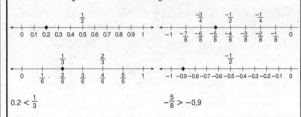

$0.2 < \frac{1}{3}$ $-\frac{5}{8} > -0.9$

Use the number lines above. Write < or >.

5. $\frac{5}{6}$ > 0.5   6. 0.6 < $\frac{2}{3}$ 7. -0.4 < $-\frac{1}{4}$ 8. $-\frac{7}{8}$ < -0.8

LESSON 2-2 Challenge
From Repeating Decimal to Fraction

You can use an equation to write a repeating decimal as a fraction.

Write 0.3333... as a fraction.   Write 0.363636... as a fraction.
Let $x = 0.3333...$ Let $x = 0.363636...$
Then $10x = 3.3333...$ Then $100x = 36.363636...$

$10x = 3.3333...$ $100x = 36.3636...$
$\underline{-x = 0.3333...}$ $\underline{-x = 0.3636...}$
$9x = 3$ Subtract. $99x = 36$ Subtract.
$\frac{9x}{9} = \frac{3}{9}$ Solve the equation. $\frac{99x}{99} = \frac{36}{99}$ Solve the equation.
$x = \frac{1}{3}$ Simplify. $x = \frac{4}{11}$ Simplify.
So, $0.3333... = \frac{1}{3}$.   So, $0.3636... = \frac{4}{11}$.

Write each repeating decimal as a fraction.

1. $0.6666... = \frac{2}{3}$ 2. $0.8888... = \frac{8}{9}$

3. $0.454545... = \frac{5}{11}$ 4. $0.090909... = \frac{1}{11}$

5. $0.636363... = \frac{7}{11}$ 6. $0.16666... = \frac{1}{6}$

7. $0.2222... = \frac{2}{9}$ 8. $0.83333... = \frac{5}{6}$

9. $0.41666... = \frac{5}{12}$ 10. $0.58333... = \frac{7}{12}$

LESSON 2-2 Problem Solving
Comparing and Ordering Rational Numbers

Write the correct answer.

1. Carl Lewis won the gold medal in the long jump in four consecutive Summer Olympic games. He jumped 8.54 meters in 1984, 8.72 meters in 1988, 8.67 meters in 1992, 8.5 meters in 1996. Order the length of his winning jumps from least to greatest.

   <u>8.5 m, 8.54 m, 8.67 m, 8.72 m</u>

2. Scientists aboard a submarine are gathering data at an elevation of $-42\frac{1}{2}$ feet. Scientists aboard a submersible are taking photographs at an elevation of $-45\frac{1}{3}$ feet. Which scientists are closer to the surface of the ocean?

   <u>scientists aboard the submarine</u>

3. The depth of a lake is measured at three different points. Point A is -15.8 meters, Point B is -17.3 meters, and Point C is -16.9 meters. Which point has the greatest depth?

   <u>Point B</u>

4. At a swimming meet, Gail's time in her first heat was $42\frac{3}{8}$ seconds. Her time in the second heat was 42.25 seconds. Which heat did she swim faster?

   <u>the second heat</u>

The table shows the top times in a 5 K race. Choose the letter of the best answer.

5. Who had the fastest time in the race?
   A Marshall
   (B) Renzo
   C Dan
   D Aaron

Name	Time (minutes)
Marshall	18.09
Renzo	17.38
Dan	17.9
Aaron	18.61

6. Which is the slowest time in the table?
   F 18.09 minutes
   G 17.38
   H 17.9 minutes
   (J) 18.61 minutes

7. Aaron's time in a previous race was less than his time in this race but greater than Marshall's time in this race. How fast could Aaron have run in the previous race?
   A 19.24 min   (C) 18.35 min
   B 18.7 min   D 18.05 mi

LESSON 2-2 Reading Strategies
Identify Relationships

$0, \frac{1}{2},$ and 1 are common benchmarks for fractions. Sometimes you can use these benchmarks to compare fractions.

A fraction is close to 0 if its numerator is small compared to its denominator. Examples are $\frac{2}{11}, \frac{3}{20},$ and $\frac{4}{25}$.

A fraction is close to $\frac{1}{2}$ if its denominator is about twice as great as its numerator. Examples are $\frac{5}{11}, \frac{8}{15},$ and $\frac{10}{21}$.

A fraction is close to 1 if its numerator and denominator are close in value. Examples are $\frac{9}{10}, \frac{13}{15},$ and $\frac{17}{33}$.

To compare $\frac{7}{15}$ and $\frac{3}{22}$, look at the relationship between the numerator and denominator of each fraction.

   15 is a little more than 2×7, so $\frac{1}{2}$ is a benchmark for $\frac{7}{15}$.

   3 is much smaller than 22, so 0 is a benchmark for $\frac{3}{22}$.

Since $\frac{1}{2} > 0, \frac{7}{15} > \frac{3}{22}$.

Answer each question.

1. What is a benchmark for $\frac{35}{67}$? $\frac{1}{2}$

2. What is a benchmark for $\frac{11}{13}$?   1

3. Use > or < to compare $\frac{35}{67}$ and $\frac{11}{13}$. $\frac{35}{67} < \frac{11}{13}$

4. What is a benchmark for $\frac{17}{20}$?   1

5. What is a benchmark for $\frac{6}{35}$?   0

6. Use > or < to compare $\frac{17}{20}$ and $\frac{6}{35}$. $\frac{17}{20} > \frac{6}{35}$

7. Use benchmarks to compare $\frac{21}{40}$ and $\frac{19}{21}$. Explain your thinking.
   $\frac{21}{40}$ is close to $\frac{1}{2}$ and $\frac{19}{21}$ is close to 1. $\frac{1}{2} < 1$, so $\frac{21}{40} < \frac{19}{21}$

LESSON 2-2 Puzzles, Twisters, and Teasers
Rational Riddle

Why did Francis Fraction want to become a psychologist?
To answer the riddle, write the numbers in order from least to greatest.
Then write the corresponding letters in the same order.

$-1\frac{1}{4}$	A	-2.7	S
-2.05	E	$-2\frac{1}{2}$	H
$-\frac{1}{5}$	T	-2.05	E
$\frac{3}{4}$	N	-1.3	W
0	I	$-1\frac{1}{4}$	A
-2.7	S	$-1\frac{1}{8}$	S
$-\frac{5}{6}$	R	$-\frac{5}{6}$	R
$-1\frac{1}{8}$	S	-0.6	A
0.95	L	$-\frac{1}{5}$	T
-0.6	A	0	I
-1.3	W	$\frac{1}{8}$	O
0.8	A	$\frac{3}{4}$	N
$-2\frac{1}{2}$	H	0.8	A
$\frac{1}{8}$	O	0.95	L

Answer: __She was rational.__

LESSON 2-3 Practice A
Adding and Subtracting Rational Numbers

1. A statue $8\frac{5}{16}$ in. high rests on a stand that is $1\frac{3}{16}$ in. high. What is the total height?

   $9\frac{1}{2}$ in.

2. During the 19th Olympic Winter Games in 2002, the United States 4-man bobsled teams won silver and bronze medals. USA-1 sled had a total time of 3 min 7.81 sec. The USA-2 sled had a total time of 3 min 7.86 sec. What is the difference in the time of the two runs?

   0.05 sec

Use a number line to find each sum.

3. $-0.2 + 0.6$

   0.4

4. $\frac{1}{5} + \frac{3}{5}$

   $\frac{4}{5}$

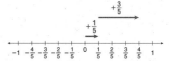

Add or subtract. Write each answer in simplest form.

5. $\frac{2}{9} + \frac{4}{9}$ = $\frac{2}{3}$

6. $\frac{5}{12} + \frac{3}{12}$ = $\frac{2}{3}$

7. $\frac{9}{10} - \frac{7}{10}$ = $\frac{1}{5}$

8. $\frac{8}{15} - \frac{11}{15}$ = $-\frac{1}{5}$

9. $\frac{3}{14} - \frac{9}{14}$ = $-\frac{3}{7}$

10. $\frac{5}{18} - \frac{11}{18}$ = $-\frac{1}{3}$

11. $\frac{1}{8} + \frac{5}{8}$ = $\frac{3}{4}$

12. $\frac{5}{6} + \frac{1}{6}$ = 1

Evaluate each expression for the given value of the variable.

13. $18.3 + x$ for $x = -1.6$

    16.7

14. $20.6 + x$ for $x = 2.8$

    23.4

15. $\frac{9}{11} + x$ for $x = -\frac{5}{11}$

    $\frac{4}{11}$

LESSON 2-3 Practice B
Adding and Subtracting Rational Numbers

1. Gretchen bought a sweater for $23.89. In addition, she had to pay $1.43 in sales tax. She gave the sales clerk $30. How much change did Gretchen receive from her total purchase?

   $4.68

2. Jacob is replacing the molding around two sides of a picture frame. The measurements of the sides of the frame are $4\frac{3}{16}$ in. and $2\frac{5}{16}$ in. What length of molding will Jacob need?

   $6\frac{1}{2}$ in.

Use a number line to find each sum.

3. $-0.5 + 0.4$

   -0.1

4. $-\frac{2}{7} + \frac{6}{7}$

   $\frac{4}{7}$

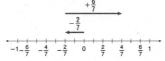

Add or subtract. Simplify.

5. $\frac{3}{8} + \frac{1}{8}$ = $\frac{1}{2}$

6. $-\frac{1}{10} + \frac{7}{10}$ = $\frac{3}{5}$

7. $\frac{5}{14} - \frac{3}{14}$ = $\frac{1}{7}$

8. $\frac{4}{15} + \frac{7}{15}$ = $\frac{11}{15}$

9. $\frac{5}{18} - \frac{7}{18}$ = $-\frac{1}{9}$

10. $-\frac{8}{17} - \frac{2}{17}$ = $-\frac{10}{17}$

11. $-\frac{1}{16} + \frac{5}{16}$ = $\frac{1}{4}$

12. $\frac{3}{20} + \frac{1}{20}$ = $\frac{1}{5}$

Evaluate each expression for the given value of the variable.

13. $38.1 + x$ for $x = -6.1$

    32

14. $18.7 + x$ for $x = 8.5$

    27.2

15. $\frac{8}{15} + x$ for $x = -\frac{4}{15}$

    $\frac{4}{15}$

LESSON 2-3 Practice C
Adding and Subtracting Rational Numbers

1. Jesse baked a pizza and cut it into 8 pieces. He ate three pieces and his two brothers ate two pieces each. In fractional form, how much of the pizza is left?

   $\frac{1}{8}$

2. The biathlon combines cross-country skiing with rifle shooting at fixed targets. A biathlon competition features races of 6.2 mi with the contestants stopping to shoot twice, a race of 12.45 mi with four shooting stops, and a four person relay race totaling 18.6 mi. How many total miles are covered in a biathlon competition?

   37.25 mi

3. The home plate in baseball is a five-sided figure with sides that measure 17 in., 12 in., $8\frac{1}{2}$ in., 12 in., and $8\frac{1}{2}$ in. What is the sum of the sides of the figure?

   58 in.

4. Chato had $2032.64 in his checking account. He wrote two checks, one for $714.53 and the other for $289.67. What was the new balance in his checking account?

   $1028.44

Add or subtract. Simplify.

5. $\frac{9}{17} + \frac{3}{17}$ = $\frac{12}{17}$

6. $-\frac{4}{25} + \frac{19}{25}$ = $\frac{3}{5}$

7. $\frac{17}{30} + \frac{7}{30}$ = $\frac{4}{5}$

8. $-\frac{13}{20} + \frac{9}{20}$ = $-\frac{1}{5}$

9. $-\frac{8}{15} - \frac{4}{15}$ = $-\frac{4}{5}$

10. $\frac{5}{24} + \frac{13}{24}$ = $\frac{3}{4}$

11. $\frac{13}{18} - \frac{7}{18}$ = $\frac{1}{3}$

12. $-\frac{33}{50} - \frac{7}{50}$ = $-\frac{4}{5}$

Evaluate each expression for the given value of the variable.

13. $102.943 + x$ for $x = 2.03$

    104.973

14. $\frac{12}{25} - x$ for $x = -\frac{13}{25}$

    1

15. $18.01 - x$ for $x = -19.26$

    37.27

LESSON 2-3 Reteach
Adding and Subtracting Rational Numbers

To add fractions that have the same denominator:
- Use the common denominator for the sum.
- Add the numerators to get the numerator of the sum.
- Write the sum in simplest form.

$$\frac{1}{8} + \frac{3}{8} = \frac{1+3}{8} = \frac{4}{8} = \frac{1}{2}$$

To subtract fractions that have the same denominator:
- Use the common denominator for the difference.
- Subtract the numerators.
  Subtraction is addition of an opposite.
- Write the difference in simplest form.

$$\frac{3}{6} - \left(-\frac{1}{6}\right) = \frac{3+1}{6} = \frac{4}{6} = \frac{2}{3}$$

Complete to add the fractions.

1. $\frac{3}{14} + \frac{4}{14} = \frac{7}{14} = \frac{1}{2}$

2. $\frac{2}{10} + \left(-\frac{4}{10}\right) = -\frac{2}{10} = -\frac{1}{5}$

3. $-\frac{5}{12} + \left(-\frac{3}{12}\right) = -\frac{8}{12} = -\frac{2}{3}$

Complete to subtract the fractions.

4. $\frac{8}{9} - \frac{2}{9} = \frac{6}{9} = \frac{2}{3}$

5. $\frac{9}{15} - \left(-\frac{3}{15}\right) = \frac{12}{15} = \frac{4}{5}$

6. $-\frac{10}{24} - \left(-\frac{2}{24}\right) = -\frac{8}{24} = -\frac{1}{3}$

To add or subtract decimals, line up the decimal points and then add or subtract from right to left as usual.

```
  12.83         35.78
 +24.17       − 14.55
  37.00         21.23
```

Complete to add the decimals.

7. $14.23 + 3.56 = \underline{17.79}$
8. $44.02 + 8.07 = \underline{52.09}$
9. $1.39 + 13.6 = \underline{14.99}$

Complete to subtract the decimals.

10. $124.33 - 13.16 = \underline{111.17}$
11. $33.47 - 0.6 = \underline{32.87}$
12. $25.15 - 25.06 = \underline{0.09}$

LESSON 2-3 Challenge
Number Code

Each sum is the code for a letter. As you find a sum, write its letter code in the message below. Write the sum in simplest form. Some letters appear more than once. An example is done for you.

$4.5 + (-6.5)$ __−2__, T   1. $14.56 + (-10.09)$ __4.47__, V
2. $\frac{7}{8} + \left(-1\frac{3}{8}\right)$ __$-\frac{1}{2}$__, M   3. $\frac{6}{8} + \left(-\frac{3}{8}\right)$ __$\frac{3}{8}$__, N
4. $-1.05 + 0.85$ __−0.2__, I   5. $\frac{-2}{4} + \left(\frac{-3}{4}\right)$ __$-1\frac{1}{4}$__, U
6. $-7.08 + (-12.02)$ __−19.1__, S   7. $-9.5 + 3.1$ __−6.4__, E
8. $\frac{-4}{5} + 1$ __$\frac{1}{5}$__, E   9. $-1\frac{1}{2} + \left(-1\frac{1}{2}\right)$ __−3__, H
10. $1\frac{2}{4} + \left(\frac{-3}{4}\right)$ __$\frac{3}{4}$__, P   11. $5 + \left(-4\frac{1}{10}\right)$ __$\frac{9}{10}$__, I
12. $8 + (-6.4)$ __1.6__, Y   13. $-3\frac{1}{4} + 3\frac{1}{4}$ __0__, S
14. $\frac{7}{8} + \left(-1\frac{7}{8}\right)$ __−1__, L   15. $6.52 + (-5)$ __1.52__, Z
16. $-62.3 + 23.9$ __−38.4__, A   17. $9\frac{1}{8} + (-10)$ __$-\frac{7}{8}$__, R
18. $-2.9 + 0.85$ __−2.05__, O   19. $2.7 + (-0.9)$ __1.8__, O

S U R E   Y O U   A R E
−19.1 −1¼ −⅞ −6.4   1.6 −2.05 −1¼   −38.4 −⅞ ⅕

N O T   L E S S   T H A N
⅜ 1.8 −1   ⅕ −19.1 0   −2 −3 −38.4 ⅜

Z E R O ?
1.52 −6.4 −⅞ −2.05

I' M   P O S I T I V E !
−0.2 −½   ¾ −2.05 0 9/10 −2 −0.2 4.47 ⅕

LESSON 2-3 Problem Solving
Adding and Subtracting Rational Numbers

Write the correct answer.

1. In 2004, Yuliya Nesterenko of Belarus won the Olympic Gold in the 100-m dash with a time of 10.93 seconds. In 2000, American Marion Jones won the 100-m dash with a time of 10.75 seconds. How many seconds faster did Marion Jones run the 100-m dash?
   __0.18 s__

2. The snowfall in Rochester, NY in the winter of 1999–2000 was 91.5 inches. Normal snowfall is about 76 inches per winter. How much more snow fell in the winter of 1999–2000 than is normal?
   __15.5 inches__

3. In a survey, $\frac{76}{100}$ people indicated that they check their e-mail daily, while $\frac{23}{100}$ check their e-mail weekly, and $\frac{1}{100}$ check their e-mail less than once a week. What fraction of people check their e-mail at least once a week?
   __$\frac{99}{100}$__

4. To make a small amount of play dough, you can mix the following ingredients: 1 cup of flour, $\frac{1}{2}$ cup of salt and $\frac{1}{2}$ cup of water. What is the total amount of ingredients added to make the play dough?
   __2 cups__

Choose the letter for the best answer.

5. How much more expensive is it to buy a ticket in Boston than in Minnesota?
   A $20.95   C $5.40
   B $55.19   **D $26.35**

Baseball Ticket Prices	
Location	Average Price
Minnesota	$14.42
League Average	$19.82
Boston	$40.77

6. How much more expensive is it to buy a ticket in Boston than the league average?
   F $60.59
   G $20.95
   H $5.40
   J $26.35

7. What is the total cost of a ticket in Boston and a ticket in Minnesota?
   A $55.19
   B $34.24
   C $60.59
   D $54.19

LESSON 2-3 Reading Strategies
Use a Visual Model

A number line can help you picture addition with decimals. This number line is divided into tenths.

Add $-0.6 + 1.8$.

1. Where do you start on the number line? __at 0__
2. Do you move to the right or to the left? Why?
   __to the left; because negative numbers are to the left of zero__
3. How many places do you move? __6 of the tenths__
4. To add, do you move to the right or to the left? __to the right__
5. How many places do you move? __18 of the tenths__
6. At what number do you end? __1.2__

This number line helps you picture addition with fractions. The number line is divided into sixths.

Add $-\frac{4}{6} + 1\frac{3}{6}$.

7. Do you move first to the left or right from 0 on the number line? Why?
   __to the left, because negative numbers are to the left of 0__
8. How many places do you move? __4 of the sixths__
9. To add, do you move to the right or left? __to the right__
10. How many places to you move? __9 of the sixths__
11. At what number do you end? __$\frac{5}{6}$__

LESSON 2-3 Puzzles, Twisters & Teasers
Bee a Math Master!

Add or subtract to find the answers. Then solve the riddle using the letters associated with the answers.

S $-\frac{1}{12} + (-\frac{7}{12}) =$ __$-\frac{2}{3}$__

F $-0.9 + 2.5 =$ __1.6__

D $\frac{8}{11} - \frac{3}{11} =$ __$\frac{5}{11}$__

O $-\frac{4}{13} - \frac{8}{13} =$ __$-\frac{12}{13}$__

R $-\frac{1}{15} + \frac{13}{15} =$ __$\frac{4}{5}$__

G $\frac{11}{32} - \frac{27}{32} =$ __$-\frac{1}{2}$__

W $-0.06 + 0.86 =$ __0.8__

T $-\frac{19}{25} + \frac{13}{25} =$ __$-\frac{6}{25}$__

E $\frac{8}{21} + \frac{15}{21} =$ __$\frac{23}{21}$__

H $0.9 + 0.3 =$ __1.2__

Why did the bee hum?

It F   O   R   G   O   T   T   H   E
   1.6 -$\frac{12}{13}$ $\frac{4}{5}$ -$\frac{1}{2}$ -$\frac{12}{13}$ -$\frac{6}{25}$ -$\frac{6}{25}$ 1.2 $\frac{23}{21}$

   W   O   R   D   S
   0.8 -$\frac{12}{13}$ $\frac{4}{5}$ $\frac{5}{11}$ -$\frac{2}{3}$

LESSON 2-4 Practice A
Multiplying Rational Numbers

Multiply. Write each answer in simplest form.

1. $5(\frac{1}{3})$ $\frac{5}{3}$ or $1\frac{2}{3}$
2. $-2(\frac{2}{5})$ $-\frac{4}{5}$
3. $4(\frac{1}{6})$ $\frac{2}{3}$
4. $-3(\frac{2}{9})$ $-\frac{2}{3}$

5. $-\frac{5}{7}(\frac{2}{5})$ $-\frac{2}{7}$
6. $\frac{3}{4}(\frac{1}{3})$ $\frac{1}{4}$
7. $-\frac{1}{4}(\frac{1}{3})$ $-\frac{1}{12}$
8. $-\frac{1}{6}(-\frac{2}{3})$ $\frac{1}{9}$

9. $\frac{1}{2}(\frac{10}{7})$ $\frac{5}{7}$
10. $\frac{3}{10}(-\frac{5}{18})$ $-\frac{1}{12}$
11. $\frac{4}{5}(-\frac{12}{16})$ $-\frac{3}{5}$
12. $\frac{4}{3}(\frac{24}{16})$ 2

13. $-4(1\frac{1}{2})$ -6
14. $\frac{3}{4}(\frac{5}{8})$ $\frac{15}{32}$
15. $-\frac{2}{5}(3\frac{1}{4})$ $-1\frac{3}{10}$
16. $-\frac{5}{6}(-\frac{3}{10})$ $\frac{1}{4}$

Multiply.

17. -3.2×5 = -16
18. 0.34×0.06 = 0.0204
19. -8.12×-9 = 73.08
20. 4.24×3.5 = 14.84

21. -3.14×0.007 = -0.02198
22. -6.7×0.8 = -5.36
23. -0.25×-2.4 = 0.6
24. 7.9×-2 = -15.8

25. Jade babysat $4\frac{1}{2}$ hours for the Lenox family. She was paid $5 an hour. How much did she receive for this babysitting job?

$22.50

LESSON 2-4 Practice B
Multiplying Rational Numbers

Multiply. Write each answer in simplest form.

1. $8(\frac{3}{4})$ 6
2. $-6(\frac{9}{18})$ -3
3. $-9(\frac{5}{6})$ $-7\frac{1}{2}$
4. $-6(-\frac{7}{12})$ $3\frac{1}{2}$

5. $-\frac{5}{18}(\frac{8}{15})$ $-\frac{4}{27}$
6. $\frac{7}{12}(\frac{14}{21})$ $\frac{7}{18}$
7. $-\frac{1}{9}(\frac{27}{24})$ $-\frac{1}{8}$
8. $-\frac{1}{11}(-\frac{3}{2})$ $\frac{3}{22}$

9. $\frac{7}{20}(-\frac{15}{28})$ $-\frac{3}{16}$
10. $\frac{16}{25}(-\frac{18}{32})$ $-\frac{9}{25}$
11. $\frac{1}{9}(-\frac{18}{17})$ $-\frac{2}{17}$
12. $\frac{17}{20}(-\frac{12}{34})$ $-\frac{3}{10}$

13. $-4(2\frac{1}{6})$ $-8\frac{2}{3}$
14. $\frac{3}{4}(1\frac{3}{8})$ $1\frac{1}{32}$
15. $3\frac{1}{5}(\frac{2}{3})$ $2\frac{2}{15}$
16. $-\frac{5}{6}(2\frac{1}{2})$ $-2\frac{1}{12}$

Multiply.

17. $-2(-5.2)$ = 10.4
18. $0.53(0.04)$ = 0.0212
19. $(-7)(-3.9)$ = 27.3
20. $-2(8.13)$ = -16.26

21. $0.02(-4.62)$ = -0.0924
22. $0.5(-7.8)$ = -3.9
23. $(-0.41)(-8.5)$ = 3.485
24. $(-8)(6.3)$ = -50.4

25. $15(-0.05)$ = -0.75
26. $(-3.04)(-1.7)$ = 5.168
27. $10(-0.09)$ = -0.9
28. $(-0.8)(-0.15)$ = 0.12

29. Travis painted for $6\frac{2}{3}$ hours. He received $27 an hour for his work. How much was Travis paid for doing this painting job?

$180

LESSON 2-4 Practice C
Multiplying Rational Numbers

Multiply. Write each answer in simplest form.

1. $10(\frac{4}{5})$ 8
2. $-12(\frac{5}{24})$ $-3\frac{1}{2}$
3. $-11(\frac{5}{22})$ $-2\frac{1}{2}$
4. $-18(\frac{5}{36})$ $-2\frac{1}{2}$

5. $\frac{14}{28}(\frac{7}{42})$ $\frac{1}{12}$
6. $\frac{25}{64}(\frac{16}{75})$ $\frac{1}{12}$
7. $\frac{14}{19}(\frac{38}{70})$ $\frac{2}{5}$
8. $-\frac{5}{27}(-\frac{9}{35})$ $\frac{1}{21}$

9. $\frac{9}{20}(\frac{36}{81})$ $\frac{1}{5}$
10. $1\frac{7}{10}(-\frac{5}{17})$ $-\frac{1}{2}$
11. $\frac{39}{50}(-\frac{35}{117})$ $-\frac{7}{30}$
12. $1\frac{1}{3}(-\frac{63}{168})$ $-\frac{1}{2}$

13. $-3(2\frac{32}{64})$ $-7\frac{1}{2}$
14. $\frac{2}{38}(9\frac{1}{2})$ $\frac{1}{2}$
15. $-\frac{16}{24}(1\frac{32}{34})$ $-1\frac{1}{6}$
16. $3\frac{1}{4}(\frac{8}{48})$ $\frac{13}{24}$

Multiply.

17. $15(-13.5)$ = -202.5
18. $6.34(1.08)$ = 6.8472
19. $(-19)(-11.82)$ = 224.58
20. $8.5(16.42)$ = 139.57

21. $3.08(-4.38)$ = -13.4904
22. $2.8(7.15)$ = 20.02
23. $(-2.25)(-2.25)$ = 5.0625
24. $(16)(2.001)$ = 32.016

25. Mrs. Johnson harvested 107 pounds of tomatoes from her garden. She sold them for $0.85 a pound. How much did she receive from selling all the tomatoes?

$90.95

26. A store is having a clearance sale of $\frac{1}{3}$ off the regular price. How much will be saved on a jacket with a regular price of $154.35? What will be the sale price of the jacket?

$51.45; $102.90

Copyright © by Holt, Rinehart and Winston. 77 Holt Mathematics

Reteach
2-4 Multiplying Rational Numbers

To model $\frac{1}{3} \times \frac{3}{4}$:

Divide a square into 4 equal parts. Lightly shade 3 of the 4.	Darken 1 of the 3 shaded parts.	Compare the 1 darkened part to the original 4.

$$\frac{1}{3} \times \frac{3}{4} = \frac{1}{4}$$

Model each multiplication. Write the result. Possible models are shown.

1. $\frac{1}{2} \times \frac{2}{4} = \underline{\frac{1}{4}}$

2. $\frac{3}{4} \times \frac{4}{6} = \underline{\frac{1}{2}}$

3. $\frac{2}{3} \times \frac{3}{9} = \underline{\frac{2}{9}}$

To multiply fractions:
- Cancel common factors, one in a numerator and the other in a denominator.
- Multiply the remaining factors in the numerator and in the denominator.
- If the signs of the factors are the same, the product is positive. If the signs of the factors are different, the product is negative.

$\frac{3}{4} \times \frac{8}{9} = \frac{1 \times 2}{1 \times 3} = \frac{2}{3}$

Multiply. Answer in simplest form.

4. $\frac{1}{2} \times \frac{4}{9} = \underline{\frac{2}{9}}$
5. $\frac{2}{3} \times \frac{6}{7} = \underline{\frac{4}{7}}$
6. $\frac{3}{5} \times \frac{15}{17} = \underline{\frac{9}{17}}$

7. $\frac{2}{3} \times \left(-\frac{9}{10}\right) = \underline{-\frac{3}{5}}$
8. $\left(-\frac{2}{9}\right) \times \frac{27}{40} = \underline{-\frac{3}{20}}$
9. $\left(-\frac{4}{7}\right) \times \left(-\frac{21}{8}\right) = \underline{1\frac{1}{2}}$

Challenge
2-4 Curtains

Variations of modern long multiplication were introduced into Europe by a 13th-century Italian, Leonardo of Pisa (Fibonacci). Many multiplication techniques can be traced to a book called *Lilaviti*, written by Bhaskara for his daughter in 12th-century India.

Here's how to do multiplication by the **Gelosia Method**, named after *jalousie*, the iron grill Italians placed over the windows. The method is also called the **Lattice Method of Multiplication**.

Consider: 38 × 56

Divide a square as shown.	Align the factors. Insert the individual products.	Sum each diagonal; begin with lower right. As needed, carry numbers into next diagonal sum.
		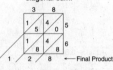

So, 38 × 56 = 2128.

Use the Gelosia Method to multiply. Verify results by your usual method.

1.
2.
3.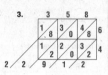

39 × 47 = __1833__   68 × 73 = __4964__   358 × 64 = __22,912__

Problem Solving
2-4 Multiplying Rational Numbers

Use the table at the right.

Average World Births and Deaths per Second in 2001	
Births	$4\frac{1}{5}$
Deaths	1.7

1. What was the average number of births per minute in 2001?
   __252 births__

2. What was the average number of deaths per hour in 2001?
   __6,120 deaths__

3. What was the average number of births per day in 2001?
   __362,880 births__

4. What was the average number of births in $\frac{1}{2}$ of a second in 2001?
   __$2\frac{1}{10}$ births__

5. What was the average number of births in $\frac{1}{4}$ of a second in 2001?
   __$1\frac{1}{20}$ births__

Use the table below. During exercise, the target heart rate is 0.5–0.75 of the maximum heart rate. Choose the letter for the best answer.

Age	Maximum Heart Rate
13	207
14	206
15	205
20	200
25	195

Source: American Heart Association

6. What is the target heart rate range for a 14 year old?
   A 7–10.5
   B 103–154.5
   C 145–166
   D 206–255

7. What is the target heart rate range for a 20 year old?
   F 100–150
   G 125–175
   H 150–200
   J 200–250

8. What is the target heart rate range for a 25 year old?
   A 25–75
   B 85–125
   C 97.5–146.25
   D 195–250

Reading Strategies
2-4 Use a Visual Model

This rectangle will help you understand how to find the product of $\frac{1}{2} \cdot \frac{1}{3}$. First, $\frac{1}{2}$ of the rectangle was shaded. Then, the rectangle was divided horizontally into thirds. Then, $\frac{1}{3}$ was shaded. The overlap of the shading shows the product of $\frac{1}{2} \cdot \frac{1}{3}$.

1. Into how many parts is the rectangle divided? What fractional part of the rectangle is each of these parts? __6; $\frac{1}{6}$__

2. What fractional part of the rectangle has shading that overlaps? __$\frac{1}{6}$__

3. Multiply the numerators and the denominators of the given fractions. __$\frac{1}{2} \cdot \frac{1}{3} = \frac{1}{6}$__

4. Use the rectangle to draw a model for the problem $\frac{1}{4} \cdot \frac{1}{2}$.

5. Draw lines from top to bottom to divide the rectangle into fourths. Shade one-fourth of the rectangle.

6. Draw a line across the rectangle to divide it into halves. Into how many parts is the rectangle now divided? __8__

7. Shade one of the halves.

8. What fractional part of the rectangle was shaded twice? __$\frac{1}{8}$__

9. Multiply the numerators and denominators. __$\frac{1}{4} \cdot \frac{1}{2} = \frac{1}{8}$__

LESSON 2-4 Puzzles, Twisters & Teasers
Egg-zactly Correct!

Multiply. Write each answer in its simplest form. Then solve the riddle using the answers.

Y $\frac{7}{8}\left(\frac{3}{5}\right) =$ __$\frac{21}{40}$__

N $3\left(2\frac{1}{5}\right) =$ __$6\frac{3}{5}$__

T $-2\left(\frac{9}{16}\right) =$ __$-1\frac{1}{8}$__

E $6\left(\frac{2}{3}\right) =$ __4__

I $-5\left(1\frac{3}{4}\right) =$ __$-8\frac{3}{4}$__

W $2\left(\frac{7}{8}\right) =$ __$1\frac{3}{4}$__

C $\frac{3}{4}\left(-\frac{1}{8}\right) =$ __$-\frac{3}{32}$__

O $\frac{6}{8}\left(\frac{2}{5}\right) =$ __$\frac{3}{10}$__

K $1\frac{2}{3}\left(\frac{5}{6}\right) =$ __$1\frac{7}{18}$__

L $-\frac{1}{3}\left(-\frac{4}{7}\right) =$ __$\frac{4}{21}$__

What do you call a city with a million eggs?

N E W   Y O L K
$6\frac{3}{5}$ 4 $1\frac{3}{4}$ $\frac{21}{40}$ $\frac{3}{10}$ $\frac{4}{21}$ $1\frac{7}{18}$

C I T Y
$-\frac{3}{32}$ $-8\frac{3}{4}$ $-1\frac{1}{8}$ $\frac{21}{40}$

LESSON 2-5 Practice A
Dividing Rational Numbers

Divide. Write each answer in simplest form.

1. $\frac{1}{4} \div \frac{3}{8}$ __$\frac{2}{3}$__
2. $-\frac{2}{3} \div \frac{5}{9}$ __$-1\frac{1}{5}$__
3. $\frac{1}{6} \div \frac{1}{3}$ __$\frac{1}{2}$__
4. $\frac{3}{4} \div \left(-\frac{1}{8}\right)$ __-6__
5. $\frac{1}{9} \div \frac{1}{3}$ __$\frac{1}{3}$__
6. $\frac{2}{5} \div \frac{4}{7}$ __$\frac{7}{10}$__
7. $-\frac{3}{5} \div \frac{6}{7}$ __$-\frac{7}{10}$__
8. $-\frac{3}{8} \div \left(-\frac{5}{6}\right)$ __$\frac{9}{20}$__
9. $1\frac{2}{5} \div 1\frac{1}{2}$ __$\frac{14}{15}$__
10. $-\frac{3}{4} \div 9$ __$-\frac{1}{12}$__
11. $-2\frac{1}{3} \div \frac{1}{4}$ __$-9\frac{1}{3}$__
12. $-\frac{5}{8} \div 5$ __$-\frac{1}{8}$__

Divide.

13. $1.53 \div 0.3$ __5.1__
14. $5.14 \div 0.2$ __25.7__
15. $10.05 \div 0.05$ __201__
16. $5.28 \div 0.4$ __13.2__
17. $6.54 \div 0.03$ __218__
18. $29.45 \div 0.005$ __5,890__
19. $8.58 \div 0.06$ __143__
20. $1.61 \div 0.7$ __2.3__

Evaluate each expression for the given value of the variable.

21. $\frac{10}{x}$ for $x = 0.05$ __200__
22. $\frac{9.12}{x}$ for $x = -0.2$ __-45.6__
23. $\frac{42.42}{x}$ for $x = 1.4$ __30.3__

24. Mr. Chen has a 76-in. space to stack books. Each book is $6\frac{1}{3}$ in. tall. How many books can he stack in the space?
__12 books__

LESSON 2-5 Practice B
Dividing Rational Numbers

Divide. Write each answer in simplest form.

1. $\frac{1}{5} \div \frac{3}{10}$ __$\frac{2}{3}$__
2. $-\frac{5}{8} \div \frac{3}{4}$ __$-\frac{5}{6}$__
3. $\frac{1}{4} \div \frac{1}{8}$ __2__
4. $-\frac{2}{3} \div \frac{4}{15}$ __$-2\frac{1}{2}$__
5. $1\frac{2}{9} \div 1\frac{2}{3}$ __$\frac{11}{15}$__
6. $-\frac{7}{10} \div \left(\frac{2}{5}\right)$ __$-1\frac{3}{4}$__
7. $\frac{6}{11} \div \frac{3}{22}$ __4__
8. $\frac{4}{9} \div \left(-\frac{8}{15}\right)$ __$-\frac{5}{6}$__
9. $\frac{3}{8} \div -15$ __$-\frac{1}{40}$__
10. $-\frac{5}{6} \div 12$ __$-\frac{5}{72}$__
11. $6\frac{1}{2} \div 1\frac{5}{8}$ __4__
12. $-\frac{9}{10} \div 6$ __$-\frac{3}{20}$__

Divide.

13. $24.35 \div 0.5$ __48.7__
14. $2.16 \div 0.04$ __54__
15. $3.16 \div 0.02$ __158__
16. $7.32 \div 0.3$ __24.4__
17. $87.36 \div 0.6$ __145.6__
18. $79.36 \div 0.8$ __99.2__
19. $4.27 \div 0.007$ __610__
20. $63.81 \div 0.9$ __70.9__
21. $1.23 \div 0.003$ __410__
22. $62.46 \div 0.09$ __694__
23. $21.12 \div 0.4$ __52.8__
24. $82.68 \div 0.06$ __1378__

Evaluate each expression for the given value of the variable.

25. $\frac{18}{x}$ for $x = 0.12$ __150__
26. $\frac{10.8}{x}$ for $x = 0.03$ __360__
27. $\frac{9.18}{x}$ for $x = -1.2$ __-7.65__

28. A can of fruit contains $3\frac{1}{2}$ cups of fruit. The suggested serving size is $\frac{1}{2}$ cup. How many servings are in the can of fruit?
__7 servings__

LESSON 2-5 Practice C
Dividing Rational Numbers

Divide. Write each answer in simplest form.

1. $\frac{10}{15} \div \frac{8}{25}$ __$2\frac{1}{12}$__
2. $1\frac{3}{18} \div 2\frac{1}{3}$ __$\frac{1}{2}$__
3. $\frac{6}{13} \div \frac{18}{26}$ __$\frac{2}{3}$__
4. $-\frac{12}{21} \div \frac{3}{14}$ __$-2\frac{2}{3}$__
5. $-\frac{3}{20} \div \left(\frac{15}{35}\right)$ __$-\frac{7}{20}$__
6. $-\frac{27}{32} \div 3$ __$-\frac{9}{32}$__
7. $\frac{36}{41} \div \frac{9}{82}$ __8__
8. $2\frac{12}{34} \div 1\frac{3}{17}$ __2__

Divide.

9. $3.15 \div 0.05$ __63__
10. $26.008 \div 0.4$ __65.02__
11. $-983.1 \div (-0.3)$ __3277__
12. $1.44 \div 0.16$ __9__
13. $236.4 \div 0.0012$ __197,000__
14. $-10.08 \div 0.005$ __-2016__
15. $2.253 \div 0.15$ __15.02__
16. $-1.161 \div 0.18$ __-6.45__

Evaluate each expression for the given value of the variable.

17. $\frac{2.25}{x}$ for $x = -0.009$ __-250__
18. $\frac{-234.72}{x}$ for $x = 3.6$ __-65.2__
19. $\frac{-4330.8}{x}$ for $x = -8.02$ __540__

20. Emma bought $2\frac{1}{2}$ yards of cording for the trim around the edge of a square pillow. How much will she use for each side of the pillow?
__$\frac{5}{8}$ yard__

21. Sean has a loan of $8804.46 including interest. He makes payments of $209.63 each month on the simple interest loan. How many months will it take for Sean to repay his loan?
__42 months__

Holt Mathematics

LESSON 2-5 Reteach
Dividing Rational Numbers

To write the **reciprocal** of a fraction, interchange the numerator and denominator.
The product of a number and its reciprocal is 1.

$\frac{2}{3} \times \frac{3}{2} = 1$ Fraction → Reciprocal

Write the reciprocal of each rational number.

1. The reciprocal of $\frac{3}{5}$ is: $\underline{\frac{5}{3}}$
2. The reciprocal of 6 is: $\underline{\frac{1}{6}}$
3. The reciprocal of $2\frac{1}{3}$ is: $\underline{\frac{3}{7}}$

To divide by a fraction, multiply by its reciprocal.

$\frac{2}{3} \div 6$ $\frac{3}{5} \div \frac{9}{10}$
$\frac{2}{3} \times \frac{1}{6}$ $\frac{3}{5} \times \frac{10}{9}$
$\frac{1 \times 1}{3 \times 3} = \frac{1}{9}$ $\frac{3 \times 10}{5 \times 9} = \frac{2}{3}$

Complete to divide and simplify.

4. $\frac{3}{8} \div 12 = \frac{3}{8} \times \underline{\frac{1}{12}} = \underline{\frac{3}{96}} = \underline{\frac{1}{32}}$
5. $\frac{4}{3} \div 16 = \underline{\frac{4}{3}} \times \underline{\frac{1}{16}} = \underline{\frac{4}{48}} = \underline{\frac{1}{12}}$
6. $\frac{5}{7} \div \frac{20}{21} = \frac{5}{7} \times \underline{\frac{21}{20}} = \underline{\frac{3}{4}}$
7. $-\frac{3}{4} \div \left(\frac{9}{8}\right) = -\frac{3}{4} \times \underline{\left(\frac{8}{9}\right)} = \underline{-\frac{2}{3}}$

Change a decimal divisor to a whole number. Using the number of places in the divisor, move the decimal point to the right in both the divisor and the dividend.

$0.7\overline{)4.34} \rightarrow 0.7\overline{)4.3.4} \rightarrow 7\overline{)43.4}$ = 6.2

Rewrite each division with a whole-number divisor. Then, do the division.

8. $0.6\overline{)1.14} \rightarrow \underline{6\overline{)11.4}} = \underline{1.9}$
9. $0.3\overline{)4.56} \rightarrow \underline{3\overline{)45.6}} = \underline{15.2}$
10. $0.02\overline{)7.12} \rightarrow \underline{2\overline{)712}} = \underline{356}$
11. $0.08\overline{)57.28} \rightarrow \underline{8\overline{)5728}} = \underline{716}$

LESSON 2-5 Challenge
A New License to Operate

You can invent new operations based on the familiar operations of addition, subtraction, multiplication, and division, and the familiar order of operations.

If $a \triangle b = \frac{a+b}{2}$ where a and b represent any rational numbers, then $3 \triangle 5 = \frac{3+5}{2} = 4$.

Use the given definition of operation △ to evaluate each expression.

1. $\frac{1}{2} \triangle (-10) = \underline{-4.75}$
2. $\frac{100 \triangle (-10)}{10} = \underline{4.5}$
3. $4 \triangle 6 \triangle 3 = \underline{4}$
4. $[5.5 \triangle (-6)] + [-6 \triangle 5.5] = \underline{-0.5}$

Use the operation shown to answer each question. $\boxed{\begin{array}{c|c} a & b \\ \hline c & d \end{array}} = ac - bd$

5. $\boxed{\begin{array}{c|c} 1 & 8 \\ \hline 3 & 4 \end{array}} = \underline{-29}$
6. $\boxed{\begin{array}{c|c} -2 & 3 \\ \hline 3 & -2 \end{array}} = \underline{0}$
7. If $\boxed{\begin{array}{c|c} 1 & 3 \\ \hline x & 2 \end{array}} = 18$, then $x = \underline{24}$
8. If $\boxed{\begin{array}{c|c} 6 & 2 \\ \hline x & x \end{array}} = 12$, then $x = \underline{3}$

Use the operation shown to answer each question. $a \diamond b = \frac{a^2}{b^2}$

9. $(3 \diamond 5) = \underline{\frac{9}{25}}$
10. $(1 \diamond 8) - (5 \diamond 8) = \underline{-\frac{3}{8}}$
11. $(1 \diamond 3) \times (3 \diamond 6) = \underline{\frac{1}{36}}$
12. $(1 \diamond 10)^2 = \underline{\frac{1}{10,000}}$

If $\lfloor n \rfloor$ means 1 less than the number of digits in the integer n, then, for example, $\lfloor 77 \rfloor = 1$ since 77 has 2 digits.

Use the definition of $\lfloor n \rfloor$ to answer each question.

13. If n is a positive integer less than 100, what is the greatest value for $\lfloor n \rfloor$? $\underline{1}$
14. If n is a positive integer less than 1001, what is the greatest value for $\lfloor n \rfloor$? $\underline{3}$
15. If n has 100 digits, what is the value of $\lfloor \lfloor n \rfloor \rfloor$? Explain.
    By definition, $\lfloor\text{100-digit number}\rfloor = 99$. Then, since 99 has 2 digits, $\lfloor 99 \rfloor = 1$.

LESSON 2-5 Problem Solving
Dividing Rational Numbers

Use the table at the right that shows the maximum speed over a quarter mile of different animals. Find the time it takes each animal to travel one-quarter mile at top speed. Round to the nearest thousandth.

1. Quarter horse — 0.005 hours
2. Greyhound — 0.006 hours
3. Human — 0.009 hours
4. Giant tortoise — 1.471 hours
5. Three-toed sloth — 1.667 hours

Maximum Speeds of Animals

Animal	Speed (mph)
Quarter Horse	47.50
Greyhound	39.35
Human	27.89
Giant Tortoise	0.17
Three-toed sloth	0.15

Choose the letter for the best answer.

6. A piece of ribbon is $1\frac{7}{8}$ inches long. If the ribbon is going to be divided into 15 pieces, how long should each piece be?
   Ⓐ $\frac{1}{8}$ in.
   B $\frac{1}{15}$ in.
   C $\frac{2}{5}$ in.
   D $28\frac{1}{8}$ in.

7. The recorded rainfall for each day of a week was 0 in., $\frac{1}{4}$ in., $\frac{3}{4}$ in., 1 in., 0 in., $1\frac{1}{4}$ in., $1\frac{1}{4}$ in. What was the average rainfall per day?
   F $\frac{9}{10}$ in.
   Ⓖ $\frac{9}{14}$ in.
   H $\frac{7}{8}$ in.
   J $4\frac{1}{2}$ in.

8. A drill bit that is $\frac{7}{32}$ in. means that the hole the bit makes has a diameter of $\frac{7}{32}$ in. Since the radius is half of the diameter, what is the radius of a hole drilled by a $\frac{7}{32}$ in. bit?
   A $\frac{14}{32}$ in.
   B $\frac{7}{32}$ in.
   C $\frac{9}{16}$ in.
   Ⓓ $\frac{7}{64}$ in.

9. A serving of a certain kind of cereal is $\frac{2}{3}$ cup. There are 12 cups of cereal in the box. How many servings of cereal are in the box?
   Ⓕ 18
   G 15
   H 8
   J 6

LESSON 2-5 Reading Strategies
Focus on Vocabulary

The word **reciprocal** means an exchange. When two friends exchange gifts, you might think of the gifts as "switching places." In the reciprocal of a fraction, the numerator and denominator exchange places.

Fraction   Reciprocal
$\frac{2}{3}$ $\frac{3}{2}$
$\frac{4}{5}$ $\frac{5}{4}$
$\frac{8}{1}$ $\frac{1}{8}$

1. What does the word *reciprocal* mean? $\underline{\text{an exchange}}$
2. What is the reciprocal of $\frac{7}{8}$? $\underline{\frac{8}{7}}$
3. What is the reciprocal of $\frac{6}{5}$? $\underline{\frac{5}{6}}$

The product of a fraction and its reciprocal is always 1.

Fraction • Reciprocal = Product

$\frac{2}{3} \cdot \frac{3}{2} = \frac{6}{6} = 1$
$\frac{4}{5} \cdot \frac{5}{4} = \frac{20}{20} = 1$
$\frac{1}{8} \cdot \frac{8}{1} = \frac{8}{8} = 1$

4. What is the product of $\frac{1}{7} \cdot \frac{7}{1}$? $\underline{\frac{7}{7}, \text{ or } 1}$
5. What is the product of $\frac{2}{6}$ and its reciprocal? $\underline{1}$
6. What is the reciprocal of $\frac{1}{2}$? $\underline{\frac{2}{1}, \text{ or } 2}$
7. What is the product of $\frac{1}{2} \times 2$? $\underline{1}$

LESSON 2-5 Puzzles, Twisters & Teasers
Hungry for Knowledge!

Solve each equation. Round answers to the nearest tenth. Then use the answers to solve the riddle.

R $17.78 \div 0.7 =$ __25.4__

L $\frac{1}{2} \div \frac{1}{4} =$ __2__

E $10.08 \div 0.6 =$ __16.8__

U $24 \div \frac{3}{4} =$ __32__

N $3.72 \div 0.3 =$ __12.4__

C $10\frac{1}{2} \div 3\frac{1}{2} =$ __3__

I $14.08 \div 0.8 =$ __17.6__

H $3\frac{1}{3} \div 1\frac{1}{5} =$ __2.8__

D $9.36 \div 0.03 =$ __312__

A $6.52 \div 0.004 =$ __1630__

What are two things you can't eat for breakfast?

L U N C H   A N D
2 32 12.4 3 2.8   1630 12.4 312

D I N N E R
312 17.6 12.4 12.4 16.8 25.4

LESSON 2-6 Practice A
Adding and Subtracting with Unlike Denominators

Name a common denominator for each sum or difference. Do not solve. **Possible answers:**

1. $\frac{1}{2} + \frac{3}{4}$ — 4
2. $\frac{1}{3} + \frac{4}{9}$ — 9
3. $\frac{2}{3} - \frac{3}{8}$ — 24
4. $\frac{1}{2} - \frac{1}{6}$ — 6

Add or subtract. Write answer in simplest form.

5. $\frac{1}{5} + \frac{1}{2}$ — $\frac{7}{10}$
6. $\frac{3}{4} + \frac{5}{6}$ — $1\frac{7}{12}$
7. $\frac{7}{10} - \frac{1}{5}$ — $\frac{1}{2}$
8. $\frac{3}{14} - \frac{4}{7}$ — $-\frac{5}{14}$

9. $3\frac{1}{3} - 1\frac{1}{6}$ — $2\frac{1}{6}$
10. $\frac{1}{4} + \frac{1}{2}$ — $\frac{3}{4}$
11. $4\frac{1}{3} - 2\frac{7}{9}$ — $1\frac{5}{9}$
12. $2\frac{3}{5} + \left(-1\frac{7}{10}\right)$ — $\frac{9}{10}$

13. $\frac{2}{5} - \frac{1}{2}$ — $-\frac{1}{10}$
14. $2\frac{1}{3} + 1\frac{1}{2}$ — $3\frac{5}{6}$
15. $3\frac{1}{4} + \left(-1\frac{5}{6}\right)$ — $1\frac{5}{12}$
16. $\frac{3}{4} - \frac{11}{12}$ — $-\frac{1}{6}$

Evaluate each expression for the given value of the variable.

17. $1\frac{7}{8} + x$ for $x = -2\frac{3}{4}$ — $-\frac{7}{8}$
18. $x - \frac{2}{3}$ for $x = \frac{1}{6}$ — $-\frac{1}{2}$
19. $x - \frac{1}{2}$ for $x = \frac{7}{8}$ — $\frac{3}{8}$

20. $2\frac{2}{3} + x$ for $x = -1\frac{5}{9}$ — $1\frac{1}{9}$
21. $x - \frac{2}{5}$ for $x = \frac{9}{10}$ — $\frac{1}{2}$
22. $x - \frac{6}{7}$ for $x = \frac{1}{2}$ — $-\frac{5}{14}$

23. Mr. Martanarie bought a new lamp and lamppost for his home. The pole was $6\frac{5}{8}$ ft tall and the lamp was $1\frac{1}{4}$ ft in height. How tall were the lamp and post together?

$7\frac{7}{8}$ ft

LESSON 2-6 Practice B
Adding and Subtracting with Unlike Denominators

Add or subtract.

1. $\frac{2}{3} + \frac{1}{2}$ — $1\frac{1}{6}$
2. $\frac{3}{5} + \frac{1}{3}$ — $\frac{14}{15}$
3. $\frac{3}{4} - \frac{1}{3}$ — $\frac{5}{12}$
4. $\frac{1}{2} - \frac{5}{9}$ — $-\frac{1}{18}$

5. $\frac{5}{16} - \frac{5}{8}$ — $-\frac{5}{16}$
6. $\frac{7}{9} + \frac{5}{6}$ — $1\frac{11}{18}$
7. $\frac{7}{8} - \frac{1}{4}$ — $\frac{5}{8}$
8. $\frac{5}{6} - \frac{3}{8}$ — $\frac{11}{24}$

9. $2\frac{7}{8} + 3\frac{5}{12}$ — $6\frac{7}{24}$
10. $1\frac{2}{9} + 2\frac{1}{18}$ — $3\frac{5}{18}$
11. $3\frac{2}{3} - 1\frac{3}{5}$ — $2\frac{1}{15}$
12. $1\frac{5}{6} + \left(-2\frac{3}{4}\right)$ — $-\frac{11}{12}$

13. $8\frac{1}{3} - 3\frac{5}{9}$ — $4\frac{7}{9}$
14. $5\frac{1}{3} + 1\frac{11}{12}$ — $7\frac{1}{4}$
15. $7\frac{1}{4} + \left(-2\frac{5}{12}\right)$ — $4\frac{5}{6}$
16. $5\frac{2}{5} - 7\frac{3}{10}$ — $-1\frac{9}{10}$

Evaluate each expression for the given value of the variable.

17. $2\frac{3}{8} + x$ for $x = 1\frac{5}{6}$ — $4\frac{5}{24}$
18. $x - \frac{2}{5}$ for $x = \frac{1}{3}$ — $-\frac{1}{15}$
19. $x - \frac{3}{10}$ for $x = \frac{3}{7}$ — $\frac{9}{70}$

20. $1\frac{5}{8} + x$ for $x = -2\frac{1}{6}$ — $-\frac{13}{24}$
21. $x - \frac{3}{4}$ for $x = \frac{1}{6}$ — $-\frac{7}{12}$
22. $x - \frac{3}{10}$ for $x = \frac{1}{2}$ — $\frac{1}{5}$

23. Ana worked $6\frac{1}{2}$ h on Monday, $5\frac{3}{4}$ h on Tuesday and $7\frac{1}{6}$ h on Friday. How many total hours did she work these three days?

$19\frac{5}{12}$ h

LESSON 2-6 Practice C
Adding and Subtracting with Unlike Denominators

Add or subtract.

1. $\frac{7}{12} + \frac{5}{9}$ — $1\frac{5}{36}$
2. $\frac{7}{10} + \frac{4}{15}$ — $\frac{29}{30}$
3. $\frac{7}{8} - \frac{11}{12}$ — $-\frac{1}{24}$
4. $\frac{15}{16} - \frac{5}{32}$ — $\frac{25}{32}$

5. $1\frac{1}{18} + \left(-4\frac{15}{24}\right)$ — $-3\frac{41}{72}$
6. $8\frac{3}{20} - 3\frac{7}{30}$ — $4\frac{11}{12}$
7. $8\frac{7}{10} + \left(-6\frac{2}{25}\right)$ — $2\frac{31}{50}$
8. $5\frac{11}{15} - 7\frac{5}{6}$ — $-2\frac{1}{10}$

Evaluate each expression for the given value of the variable.

9. $10\frac{6}{25} + x$ for $x = -2\frac{3}{5}$ — $7\frac{16}{25}$
10. $x - \frac{5}{18}$ for $x = \frac{5}{6}$ — $\frac{5}{9}$

11. $1\frac{5}{21} + x$ for $x = -5\frac{6}{7}$ — $-4\frac{13}{21}$
12. $x - \frac{8}{11}$ for $x = \frac{13}{22}$ — $-\frac{3}{22}$

13. $14\frac{1}{15} + x$ for $x = -9\frac{3}{10}$ — $4\frac{23}{30}$
14. $x - \frac{7}{9}$ for $x = \frac{2}{27}$ — $-\frac{19}{27}$

15. A carpenter cuts a piece of wood that is $9\frac{5}{16}$ ft long into two pieces. One piece is $5\frac{3}{8}$ ft long. How long is the other piece?

$3\frac{15}{16}$ ft

16. Before April 9, 2001, when the U.S. Securities and Exchange Commission ordered all U.S. stock markets to report stocks in decimals, the price of stock was reported in fractions. Under the fractional reporting system, what was the change in stock price if a stock opened the day at $31\frac{1}{8}$ and closed the day at $28\frac{5}{32}$?

down $2\frac{31}{32}$

LESSON 2-6 Reteach
Adding and Subtracting with Unlike Denominators

To model $\frac{1}{2} + \frac{1}{3}$, use two rectangles of the same size and shape.

A. 1st rectangle: Shade $\frac{1}{2}$ vertically.   **B.** 2nd rectangle: Shade $\frac{1}{3}$ horizontally.

C. Separate the shaded portions into parts of equal size.   **D.** Use a new rectangle to show the sum.

$\frac{1}{2} = \frac{3}{6}$ $\frac{1}{3} = \frac{2}{6}$ $\frac{1}{2} + \frac{1}{3} = \frac{3}{6} + \frac{2}{6} = \frac{5}{6}$

Model $\frac{1}{2} + \frac{2}{5}$. Write the result. Possible model.

1.

$\frac{1}{2}$ $\frac{2}{5}$ $\frac{1}{2} + \frac{2}{5} = \frac{5}{10} + \frac{4}{10} = \frac{9}{10}$

Model $\frac{1}{3} + \frac{3}{5}$. Write the result. Possible model.

2.

$\frac{1}{3}$ $\frac{3}{5}$ $\frac{1}{3} + \frac{3}{5} = \frac{5}{15} + \frac{9}{15} = \frac{14}{15}$

LESSON 2-6 Reteach
Adding and Subtracting with Unlike Denominators (continued)

To add fractions with different denominators, first write the fractions with common denominators. To find the LCD of denominators 5 and 6, list the multiples of each.

Multiples of 5: 5, 10, 15, 20, 25, (30)
Multiples of 6: 12, 18, 24, (30)
So, the LCD of 5 and 6 is 30.

Complete to find the LCD for each set of denominators.

3. The LCD of 6 and 4 is: __12__
Multiples of 6: __6, (12)__
Multiples of 4: __4, 8, (12)__

4. The LCD of 3 and 7 is: __21__
Multiples of 3: __3, 6, 9, 12, 15, 18, (21)__
Multiples of 7: __7, 14, (21)__

To add fractions with different denominators: Add: $\frac{1}{2} + \frac{1}{3} = \frac{1 \cdot 3}{2 \cdot 3} = \frac{3}{6}$
$\frac{1 \cdot 2}{3 \cdot 2} = \frac{2}{6}$
$= \frac{5}{6}$

Complete to add fractions. Simplify.

5. $\frac{1}{4} = \frac{5}{20}$ **6.** $\frac{3}{4} = \frac{12}{16}$ **7.** $5\frac{1}{3} = 5\frac{8}{24}$
$+ \frac{3}{5} = \frac{12}{20}$ $+ \frac{5}{16} = \frac{5}{16}$ $+ 2\frac{5}{8} = 2\frac{15}{24}$
$= \frac{17}{20}$ $= \frac{17}{16} = 1\frac{1}{16}$ $= 7\frac{23}{24}$

Add or subtract fractions. Simplify.

8. $\frac{1}{4} + \frac{7}{20} = \frac{3}{5}$ **9.** $\frac{4}{9} - \frac{1}{5} = \frac{11}{45}$ **10.** $\frac{8}{15} - \frac{1}{4} = \frac{17}{60}$

LESSON 2-6 Challenge
Please Repeat That.

A decimal that repeats one digit is equivalent to a fraction with denominator 9. $0.\overline{1} = \frac{1}{9}$ $0.\overline{2} = \frac{2}{9}$ $0.\overline{5} = \frac{5}{9}$

A decimal that repeats two digits is equivalent to a fraction with denominator 99. $0.\overline{43} = \frac{43}{99}$ $0.\overline{61} = \frac{61}{99}$ $0.\overline{38} = \frac{38}{99}$

The pattern continues so that $0.\overline{681} = \frac{681}{999}$ and $0.\overline{24793} = \frac{24,793}{99,999}$.

Use a calculator to write each decimal equivalent.

1. $\frac{1}{90} = $ __$0.0\overline{1}$__   **2.** $\frac{21}{990} = $ __$0.0\overline{21}$__   **3.** $\frac{358}{9990} = $ __$0.0\overline{358}$__

Predict the decimal equivalent of each fraction. Verify your results on a calculator.

4. $\frac{4}{90} = $ __$0.0\overline{4}$__   **5.** $\frac{62}{990} = $ __$0.0\overline{62}$__   **6.** $\frac{617}{9990} = $ __$0.0\overline{617}$__

Write each fractional equivalent and simplify.

7. $0.\overline{7} = $ **8.** $0.\overline{08} = $ **9.** $0.00\overline{24} = $
$\frac{7}{9}$ $\frac{8}{90} = \frac{4}{45}$ $\frac{24}{9900} = \frac{2}{825}$

When one digit repeats but does not begin in the first decimal place, and the digit in the first place is other than 0, you must add fractions.

$0.3\overline{7} = 0.3 + 0.0\overline{7}$
$= \frac{3}{10} + \frac{7}{90}$
$= \frac{27}{90} + \frac{7}{90} = \frac{34}{90} = \frac{17}{45}$

Write each repeating decimal as the sum of two fractions. Find the sum and simplify. Verify.

10. $0.2\overline{8} = $
$\frac{2}{10} + \frac{8}{90} = \frac{26}{90} = \frac{13}{45}$

11. $0.25\overline{32} = $
$\frac{25}{100} + \frac{32}{9900} = \frac{2507}{9900}$

12. $0.1\overline{27} = $
$\frac{1}{10} + \frac{27}{990} = \frac{126}{990} = \frac{7}{55}$

13. $0.75\overline{483} = $
$\frac{75}{100} + \frac{483}{99,900} = \frac{75,408}{99,900} = \frac{6284}{8325}$

LESSON 2-6 Problem Solving
Adding and Subtracting with Unlike Denominators

Write the correct answer.

1. Nick Hysong of the United States won the Olympic gold medal in the pole vault in 2000 with a jump of 19 ft $4\frac{1}{4}$ inches, or $232\frac{1}{4}$ inches. In 1900, Irving Baxter of the United States won the pole vault with a jump of 10 ft $9\frac{7}{8}$ inches, or $129\frac{7}{8}$ inches. How much higher did Hysong vault than Baxter?

$102\frac{3}{8}$ inches

2. In the 2000 Summer Olympics, Ivan Pedroso of Cuba won the Long jump with a jump of 28 ft $\frac{3}{4}$ inches, or $336\frac{3}{4}$ inches. Alvin Kraenzlein of the United States won the long jump in 1900 with a jump of 23 ft $6\frac{7}{8}$ inches, or $282\frac{7}{8}$ inches. How much farther did Pedroso jump than Kraenzlein?

$53\frac{7}{8}$ inches

3. A recipe calls for $\frac{1}{8}$ cup of sugar and $\frac{3}{4}$ cup of brown sugar. How much total sugar is added to the recipe?

$\frac{7}{8}$ cup

4. The average snowfall in Norfolk, VA for January is $2\frac{3}{5}$ inches, February $2\frac{9}{10}$ inches, March 1 inch, and December $\frac{9}{10}$ inches. If these are the only months it typically snows, what is the average snowfall per year?

$7\frac{2}{5}$ inches

Use the table at the right that shows the average snowfall per month in Vail, Colorado.

5. What is the average annual snowfall in Vail, Colorado?

A $15\frac{13}{20}$ in.   **C** $187\frac{1}{10}$ in.
B 153 in.   **(D)** $187\frac{4}{5}$ in.

6. The peak of the skiing season is from December through March. What is the average snowfall for this period?

F $30\frac{19}{20}$ in.   **(H)** $123\frac{4}{5}$ in.
G $123\frac{3}{5}$ in.   **J** 127 in.

Average Snowfall in Vail, CO

Month	Snowfall (in.)	Month	Snowfall (in.)
Jan	$36\frac{7}{10}$	July	0
Feb	$35\frac{7}{10}$	August	0
March	$25\frac{2}{5}$	Sept	1
April	$21\frac{1}{5}$	Oct	$7\frac{4}{5}$
May	4	Nov	$29\frac{7}{10}$
June	$\frac{3}{10}$	Dec	26

Lesson 2-6 Reading Strategies
Use a Graphic Aid

It is easy to add and subtract fractions with **common denominators**.

3 eighths + 4 eighths = 7 eighths   8 ninths − 3 ninths = 5 ninths

$\frac{3}{8} + \frac{4}{8} = \frac{7}{8}$ $\frac{8}{9} - \frac{3}{9} = \frac{5}{9}$

Adding fractions with unlike denominators requires more steps. The picture below will help you understand adding fractions with unlike denominators. $\frac{1}{2} + \frac{1}{4} = ?$

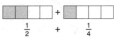

$\frac{1}{2}$ + $\frac{1}{4}$

In order to add $\frac{1}{2} + \frac{1}{4}$, you must find a common denominator.

1. What are the denominators in this problem? — 2 and 4
2. To find a common denominator, one-half can be changed into fourths. How many fourths are there in one-half? — 2
3. Change $\frac{1}{2}$ to fourths. — $\frac{1}{2} = \frac{2}{4}$
4. You can now add, because you have a common denominator. — $\frac{2}{4} + \frac{1}{4} = \frac{3}{4}$

To subtract fractions with unlike denominators, you must find a common denominator. The picture below will help you understand finding a common denominator. $\frac{5}{6} - \frac{1}{3} = ?$

$\frac{5}{6}$ $\frac{1}{3}$

5. What are the denominators in this problem? — 3, 6

To find a common denominator, you will change to sixths.

6. How many sixths are in one-third? Write the fraction. — 2; $\frac{2}{6}$
7. You can now subtract the fractions. — $\frac{5}{6} - \frac{2}{6} = \frac{3}{6}$

Lesson 2-6 Puzzles, Twisters & Teasers
Just a Tad Difficult!

Add or subtract to solve each equation. Then use the answers to solve the riddle.

M $\frac{5}{12} + \frac{3}{7} =$ $\frac{71}{84}$

T $\frac{1}{5} + \frac{7}{9} =$ $\frac{44}{45}$

O $\frac{15}{16} - \frac{9}{10} =$ $\frac{3}{80}$

H $\frac{1}{3} + 1\frac{1}{12} =$ $1\frac{5}{12}$

R $2\frac{1}{5} + 1\frac{8}{9} =$ $4\frac{4}{45}$

E $\frac{5}{8} + \frac{1}{6} =$ $\frac{19}{24}$

K $\frac{5}{16} + \frac{2}{7} =$ $\frac{67}{112}$

C $2\frac{1}{3} - 4\frac{7}{9} =$ $-2\frac{4}{9}$

A $\frac{3}{4} - \frac{5}{16} =$ $\frac{7}{16}$

N $1\frac{3}{4} + \left(-3\frac{13}{15}\right) =$ $-2\frac{7}{60}$

Where do tadpoles change into frogs?

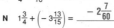

I	N	T	H	E
	$-2\frac{7}{60}$	$\frac{44}{45}$	$1\frac{5}{12}$	$\frac{19}{24}$

C	R	O	A	K	R	O	O	M
$-2\frac{4}{9}$	$4\frac{4}{45}$	$\frac{3}{80}$	$\frac{7}{16}$	$\frac{67}{112}$	$4\frac{4}{45}$	$\frac{3}{80}$	$\frac{3}{80}$	$\frac{71}{84}$

Lesson 2-7 Practice A
Solving Equations with Rational Numbers

Solve.

1. $x + 1.2 = 4.6$ $x = 3.4$
2. $a - 3.4 = 5$ $a = 8.4$
3. $2.2m = 4.4$ $m = 2$
4. $\frac{x}{1.3} = 2$ $x = 2.6$
5. $6.7 + w = -1.1$ $w = -7.8$
6. $\frac{n}{1.9} = -3.8$ $n = -7.22$
7. $7.2 = -0.9y$ $y = -8$
8. $k - 4.05 = 6.2$ $k = 10.25$
9. $\frac{d}{-3.2} = -3.75$ $d = 12$
10. $-\frac{2}{5} + x = \frac{2}{5}$ $x = \frac{4}{5}$
11. $\frac{1}{4}x = \frac{1}{2}$ $x = 2$
12. $\frac{1}{3}a = \frac{3}{4}$ $a = 2\frac{1}{4}$
13. $x - \frac{3}{2} = \frac{1}{5}$ $x = 1\frac{7}{10}$
14. $x - \frac{3}{7} = -\frac{5}{7}$ $x = -\frac{2}{7}$
15. $-\frac{5}{6}a = \frac{5}{8}$ $a = -\frac{3}{4}$

16. Elisa can reach $77\frac{3}{4}$ in. high. The ceiling is $90\frac{1}{2}$ in. high. How much higher is the ceiling than Elisa's highest reach?

    $12\frac{3}{4}$ in.

17. Nolan makes $10.60 an hour at his after-school job. Last week he worked 11.25 hr. How much was Nolan paid for the week?

    $119.25

Lesson 2-7 Practice B
Solving Equations with Rational Numbers

Solve.

1. $x + 6.8 = 12.19$ $x = 5.39$
2. $y - 10.24 = 5.3$ $y = 15.54$
3. $0.05w = 6.25$ $w = 125$
4. $\frac{a}{9.05} = 8.2$ $a = 74.21$
5. $-12.41 + x = -0.06$ $x = 12.35$
6. $\frac{d}{-8.4} = -10.2$ $d = 85.68$
7. $-2.89 = 1.7m$ $m = -1.7$
8. $n - 8.09 = -11.65$ $n = -3.56$
9. $\frac{x}{5.4} = -7.18$ $x = -38.772$
10. $\frac{7}{9} + x = 1\frac{1}{9}$ $x = \frac{1}{3}$
11. $\frac{6}{11}y = -\frac{18}{22}$ $y = -1\frac{1}{2}$
12. $\frac{7}{10}d = \frac{21}{20}$ $d = 1\frac{1}{2}$
13. $x - \left(-\frac{9}{14}\right) = \frac{5}{7}$ $x = \frac{1}{14}$
14. $x - \frac{15}{21} = 2\frac{6}{7}$ $x = 3\frac{4}{7}$
15. $-\frac{8}{15}a = \frac{9}{10}$ $a = -1\frac{11}{16}$

16. A recipe calls for $2\frac{1}{3}$ cups of flour and $1\frac{1}{4}$ cups of sugar. If the recipe is tripled, how much flour and sugar will be needed?

    7 cups of flour and $3\frac{3}{4}$ cups of sugar

17. Daniel filled the gas tank in his car with 14.6 gal of gas. He then drove 284.7 mi before needing to fill up his tank with gas again. How many miles did the car get to a gallon of gasoline?

    19.5 mi

LESSON 2-7 Practice C
Solving Equations with Rational Numbers

Solve.

1. $x + 102.8 = 89.06$
   $x = -13.74$

2. $62.5m = 2587.5$
   $m = 41.4$

3. $\frac{w}{38.7} = 51.06$
   $w = 1976.022$

4. $-10\frac{5}{18} + x = -12\frac{3}{10}$
   $x = -2\frac{1}{45}$

5. $5\frac{7}{15}a = 3\frac{2}{25}$
   $a = \frac{3}{5}$

6. $y - \left(-6\frac{1}{16}\right) = 11\frac{3}{40}$
   $y = 5\frac{1}{80}$

7. A photo that is $3\frac{1}{2}$ in. by $5\frac{1}{4}$ in. is enlarged to three times its original size. What are the new dimensions of the photo?

   $10\frac{1}{2}$ in. by $15\frac{3}{4}$ in.

8. Mars has two very small elliptical shaped moons Deimos and Phobus. They were discovered in August 1877, by Asaph Hall, an American Astronomer. The inner satellite is Phobos. It is 16.78 mi long. Deimos is the outer moon and is 9.32 mi long. What is the difference in the length of the two moons?

   7.46 mi

9. Mr. Crowley bought lunch for himself and eight of his employees. Each had a sandwich platter that costs $5.85 and a drink that costs $1.25. Five of the employees had dessert that costs $1.50. Mr. Crowley gave the delivery person $80 and told her to keep the change as a tip. How much was the delivery person's tip?

   $8.60

10. Two of the greatest rainfalls ever recorded were on July 4, 1956. In Unionville, Maryland it rained 1.23 in. in 1 min. In Curtea-de-Arges, Romania on July 7, 1889 it rained 8.1 in. in 20 min. If it had rained for 20 min in Unionville at its same record pace, what would be the difference between the two rainfall amounts?

    16.5 in.

LESSON 2-7 Reteach
Solving Equations with Rational Numbers

Solving equations with rational numbers is basically the same as solving equations with integers or whole numbers:
Use inverse operations to isolate the variable.

$\frac{1}{4}z = -16$ (Multiply each side by 4.)
$4 \cdot \frac{1}{4}z = -16 \cdot 4$
$z = -64$

$y - \frac{3}{8} = \frac{7}{8}$ (Add $\frac{3}{8}$ to each side.)
$+ \frac{3}{8} \quad + \frac{3}{8}$
$y = \frac{10}{8} = 1\frac{2}{8} = 1\frac{1}{4}$

$x + 3.5 = -17.42$ (Subtract 3.5 from each side.)
$- 3.5 \quad - 3.5$
$x = -20.92$

$-26t = 317.2$ (Divide each side by -26.)
$\frac{-26t}{-26} = \frac{317.2}{-26}$
$t = -12.2$

Tell what you would do to isolate the variable.

1. $x - 1.4 = 7.82$
   add 1.4

2. $\frac{1}{4} + y = \frac{7}{4}$
   subtract $\frac{1}{4}$

3. $3z = 5$
   divide by 3

Solve each equation.

4. $14x = -129.5$
   $x = -9.25$

5. $\frac{1}{3}y = 27$
   $y = 81$

6. $265.2 = \frac{z}{22.1}$
   $5860.92 = z$

7. $x + 53.8 = -1.2$
   $x = -55$

8. $25 = \frac{1}{5}k$
   $125 = k$

9. $m - \frac{2}{3} = \frac{3}{5}$
   $m = 1\frac{4}{15}$

LESSON 2-7 Challenge
Location, Location, Location

An equation that has a variable in the denominator of one or more of its terms is called a **fractional equation**.

One method of solution is to clear the equation of fractions by multiplying each side of the equation by the LCD.

$\frac{1}{2} + \frac{1}{x} = \frac{3}{5}$ The LCD of 2, x, and 5 is $10x$, with $x \neq 0$.

$10x\left(\frac{1}{2} + \frac{1}{x}\right) = 10x\left(\frac{3}{5}\right)$ Multiply each side by $10x$.

$10x \cdot \frac{1}{2} + 10x \cdot \frac{1}{x} = 10x \cdot \frac{3}{5}$ Distributive Property

$5x + 10 = 6x$ Simplify.

$5x - 5x + 10 = 6x - 5x$ Subtract $5x$ from each side.

$10 = x$

Check:

$\frac{1}{2} + \frac{1}{x} = \frac{3}{5}$

$\frac{1}{2} + \frac{1}{10} \stackrel{?}{=} \frac{3}{5}$ Substitute 10 for x in the original equation.

$\frac{5}{10} + \frac{1}{10} \stackrel{?}{=} \frac{3}{5}$ Do not repeat the method of solution.

$\frac{6}{10} \stackrel{?}{=} \frac{3}{5}$ Work each side separately.

$\frac{3}{5} = \frac{3}{5}$ ✓

Solve and check.

1. $\frac{4}{7} + \frac{2}{x} = \frac{2}{3}$
   $x = 21$

2. $\frac{10}{x} + \frac{8}{x} = 9$
   $x = 2$

3. $\frac{15}{x} = 7 + \frac{9}{2x}$
   $x = 1\frac{1}{2}$

LESSON 2-7 Problem Solving
Solving Equations with Rational Numbers

Write the correct answer.

1. In the last 150 years, the average height of people in industrialized nations has increased by $\frac{1}{3}$ foot. Today, American men have an average height of $5\frac{7}{12}$ feet. What was the average height of American men 150 years ago?

   $5\frac{1}{4}$ feet

2. Jaime has a length of ribbon that is $23\frac{1}{2}$ in. long. If she plans to cut the ribbon into pieces that are $\frac{3}{4}$ in. long, into how many pieces can she cut the ribbon? (She cannot use partial pieces.)

   31 pieces

3. Todd's restaurant bill for dinner was $15.55. After he left a tip, he spent a total of $18.00 on dinner. How much money did Todd leave for a tip?

   $2.45

4. The difference between the boiling point and melting point of Hydrogen is 6.47°C. The melting point of Hydrogen is -259.34°C. What is the boiling point of Hydrogen?

   -252.87°C

Choose the letter for the best answer.

5. Justin Gatlin won the Olympic gold in the 100-m dash in 2004 with a time of 9.85 seconds. His time was 0.95 seconds faster than Francis Jarvis who won the 100-m dash in 1900. What was Jarvis' time in 1900?
   A 8.95 seconds
   B 10.65 seconds
   C 10.80 seconds
   D 11.20 seconds

6. The balance in Susan's checking account was $245.35. After the bank deposited interest into the account, her balance went to $248.02. How much interest did the bank pay Susan?
   F $1.01
   G $2.67
   H $3.95
   J $493.37

7. After a morning shower, there was $\frac{17}{100}$ in. of rain in the rain gauge. It rained again an hour later and the rain gauge showed $\frac{1}{4}$ in. of rain. How much did it rain the second time?
   A $\frac{2}{25}$ in.
   B $\frac{1}{6}$ in.
   C $\frac{21}{50}$ in.
   D $\frac{3}{8}$ in.

8. Two-third of John's savings account is being saved for his college education. If $2500 of his savings is for his college education, how much money in total is in his savings account?
   F $1666.67
   G $3750
   H $4250.83
   J $5000

LESSON 2-7 Reading Strategies
Follow a Procedure

The rules for solving equations with rational numbers are the same as equations with whole numbers.

Get the variable by itself. → Perform the same operation on both sides to keep the equation balanced. → Use the rules for computing rational numbers.

Follow the steps above to help you solve $\frac{1}{4} + y = \frac{3}{4}$.

1. What is the first step to solve this equation?
   Get y by itself on one side of the equation.

2. What operation should you use?
   subtraction

3. Write an equation to show the subtraction of $\frac{1}{4}$ on both sides.
   $\frac{1}{4} - \frac{1}{4} + y = \frac{3}{4} - \frac{1}{4}$

4. What is the value of y?
   $y = \frac{2}{4}$ or $\frac{1}{2}$

Follow the steps above to solve $x - 4.5 = 13$.

5. What is the first step to solve this equation?
   Get x by itself on one side of the equation.

6. What operation should you use?
   addition

7. Write an equation to show the addition of 4.5 to both sides.
   $x - 4.5 + 4.5 = 13 + 4.5$

8. Find the value of x.
   $x = 17.5$

LESSON 2-7 Puzzles, Twisters & Teasers
Math Book Issues!

Solve the equations. Then use the letters of the variables to answer the riddle.

1. $s - \frac{5}{9} = \frac{1}{9}$ $s = \frac{2}{3}$
2. $t + \frac{1}{3} = \frac{3}{4}$ $t = \frac{5}{12}$
3. $m - \frac{1}{12} = \frac{5}{12}$ $m = \frac{1}{2}$
4. $o + \frac{5}{9} = -\frac{1}{9}$ $o = -\frac{2}{3}$
5. $e - 17.9 = 36.8$ $e = 54.7$
6. $a - 2.1 = -4.5$ $a = -2.4$
7. $l + \frac{4}{13} = \frac{12}{39}$ $l = 0$
8. $0.04n = 0.252$ $n = 6.3$
9. $b \div 3.2 = -6$ $b = -19.2$
10. $\frac{5}{9} + y = \frac{6}{18}$ $y = -\frac{2}{9}$
11. $r + \frac{5}{8} = -2\frac{3}{8}$ $r = -3$
12. $p + 3.8 = -1.6$ $p = -5.4$

Why did the student return her math book?

It had T   O   O   M   A   N   Y
       $\frac{5}{12}$ $-\frac{2}{3}$ $-\frac{2}{3}$ $\frac{1}{2}$ -2.4 6.3 $-\frac{2}{9}$

P   R   O   B   L   E   M   S
-5.4 -3 $-\frac{2}{3}$ -19.2 0 54.7 $\frac{1}{2}$ $\frac{2}{3}$

LESSON 2-8 Practice A
Solving Two-Step Equations

Describe the operation performed on both sides of the equation in steps 2 and 4.

1. $3x + 2 = 11$
   $3x + 2 - 2 = 11 - 2$ **subtract 2**
   $3x = 9$
   $\frac{3x}{3} = \frac{9}{3}$ **divide by 3**
   $x = 3$

2. $\frac{x}{4} - 1 = -2$
   $\frac{x}{4} - 1 + 1 = -2 + 1$ **add 1**
   $\frac{x}{4} = -1$
   $4\left(\frac{x}{4}\right) = 4(-1)$ **multiply by 4**
   $x = -4$

Solve.

3. $2x + 3 = 9$ $x = 3$
4. $\frac{x}{3} - 1 = 5$ $x = 18$
5. $-3a + 4 = 7$ $a = -1$
6. $\frac{x+2}{2} = -3$ $x = -8$
7. $5y - 2 = 28$ $y = 6$
8. $2x - 7 = 7$ $x = 7$
9. $\frac{w-2}{5} = -1$ $w = -3$
10. $2r + 1 = -1$ $r = -1$

Write and solve a two-step equation to answer the question.

11. Pearson rented a moving van for 1 day. The total rental charge is $66.00. A daily rental costs $45.00 plus $0.25 per mile. How many miles did he drive the van?

    $x = $ # of miles
    $45 + 0.25x = 66$
    $45 - 45 + 0.25x = 66 - 45$
    $0.25x = 21$
    $\frac{0.25x}{0.25} = \frac{21}{0.25}$
    $x = 84$ miles

LESSON 2-8 Practice B
Solving Two-Step Equations

Write and solve a two-step equation to answer the following questions.

1. The school purchased baseball equipment and uniforms for a total cost of $1762. The equipment costs $598 and the uniforms were $24.25 each. How many uniforms did the school purchase?

   $x = $ # of uniforms
   $1762 = 598 + 24.25x$
   $1762 - 598 = 598 - 598 + 24.25x$
   $1164 = 24.25x$
   $\frac{1164}{24.25} = \frac{24.25x}{24.25}$
   $48 = x$ uniforms

2. Carla runs 4 miles every day. She jogs from home to the school track, which is $\frac{3}{4}$ mile away. She then runs laps around the $\frac{1}{4}$-mile track. Carla then jogs home. How many laps does she run at the school?

   $x = $ # of laps
   $4 = \frac{3}{4} + \frac{1}{4}x + \frac{3}{4}$
   $4 - \frac{6}{4} = \frac{6}{4} - \frac{6}{4} + \frac{1}{4}x$
   $\frac{5}{2} = \frac{1}{4}x$
   $4\left(\frac{5}{2}\right) = \left(\frac{1}{4}x\right)4$
   $10 = x$ laps

Solve.

3. $\frac{a+5}{3} = 12$ $a = 31$
4. $\frac{x+2}{4} = -2$ $x = -10$
5. $\frac{y-4}{6} = -3$ $y = -14$
6. $\frac{k+1}{8} = 7$ $k = 55$
7. $0.5x - 6 = -4$ $x = 4$
8. $\frac{x}{2} + 3 = -4$ $x = -14$
9. $\frac{1}{5}n + 3 = 6$ $n = 15$
10. $2a - 7 = -9$ $a = -1$
11. $\frac{3x-1}{4} = 2$ $x = 3$
12. $-7.8 = 4.4 + 2r$ $r = -6.1$
13. $\frac{-4w+5}{-3} = -7$ $w = -4$
14. $1.3 - 5r = 7.4$ $r = -1.22$

15. A phone call costs $0.58 for the first 3 minutes and $0.15 for each additional minute. If the total charge for the call was $4.78, how many minutes was the call?   **31 minutes**

16. Seventeen less than four times a number is twenty-seven. Find the number.   **11**

LESSON 2-8 Practice C
Solving Two-Step Equations

Write an equation for each sentence, then solve it.

1. A number multiplied by five and increased by 3 is 28.
   $5x + 3 = 28; x = 5$

2. Eighteen decreased by 4 times a number is 62.
   $18 - 4x = 62; x = -11$

3. The sum of 3 times a number and 7, divided by 5, is 17.
   $\frac{3x + 7}{5} = 17; x = 26$

4. The quotient of a number and 5, minus 2, is 8.
   $\frac{x}{5} - 2 = 8; x = 50$

Solve.

5. $-13 = -3x + 14$
   $x = 9$

6. $2w - 5 = 4$
   $w = 4\frac{1}{2}$

7. $\frac{x - 5}{3} = -10$
   $x = -25$

8. $\frac{2}{3}n - 7 = 19$
   $n = 39$

9. $1.4x + 0.8 = -1.3$
   $x = -1.5$

10. $\frac{1}{4}y + \frac{3}{4} = 2$
    $y = 5$

11. $15n - 62 = -17$
    $n = 3$

12. $\frac{3}{8}a - 4 = -\frac{1}{4}$
    $a = 10$

13. $\frac{2d + 9}{6} = 11$
    $d = 28\frac{1}{2}$

14. $24.5 = 16.1 - 2.4r$
    $r = -3.5$

15. $\frac{7x - 2}{6} = -5$
    $x = -4$

16. $\frac{5}{6} - \frac{a}{4} = \frac{1}{3}$
    $a = 2$

17. Larissa is planning for a trip that cost $2145. She has $952.50 saved and is going to set aside $\frac{1}{2}$ of her weekly salary from her part-time job. Larissa earns $265 a week. How many weeks will it take her to earn the rest of the money needed for the trip?   **9 weeks**

18. Noel bought a printer for $10 less than half its original price. If Noel paid $88 for the printer, what was the original price?   **$196**

LESSON 2-8 Reteach
Solving Two-Step Equations

To solve an equation, it is important to first note how it is formed. Then, work backward to undo each operation.

$4z + 3 = 15$
The variable is multiplied by 4 and then 3 is added.
To solve, first subtract 3 and then divide by 4.

$\frac{z}{4} - 3 = 7$
The variable is divided by 4 and then 3 is subtracted.
To solve, first add 3 and then multiply by 4.

$\frac{z + 3}{4} = 7$
3 is added to the variable and then the result is divided by 4.
To solve, multiply by 4 and then subtract 3.

Describe how each equation is formed. Then, tell the steps needed to solve.

1. $3x - 5 = 7$
   The variable is __multiplied by 3__ and then __5 is subtracted__.
   To solve, first __add 5__ and then __divide by 3__.

2. $\frac{x}{3} + 5 = 7$
   The variable is __divided by 3__ and then __5 is added__.
   To solve, first __subtract 5__ and then __multiply by 3__.

3. $\frac{x + 5}{3} = 7$
   5 is __added to the variable__ and then the result is __divided by 3__.
   To solve, first __multiply by 3__ and then __subtract 5__.

4. $10 = -3x - 2$
   The variable is __multiplied by -3__ and then __2 is subtracted__.
   To solve, first __add 2__ and then __divide by -3__.

5. $10 = \frac{x - 2}{5}$
   2 is __subtracted from__ the variable and then the result is __divided by 5__.
   To solve, first __multiply by 5__ and then __add 2__.

LESSON 2-8 Reteach
Solving Two-Step Equations (continued)

To isolate the variable, work backward using inverse operations.

The variable is multiplied by 2 and then 3 is added.
$2x + 3 = 11$ To undo addition,
$-3 \quad -3$ subtract 3.
$2x = 8$ To undo multiplication,
$\frac{2x}{2} = \frac{8}{2}$ divide by 2.
$x = 4$
Check: Substitute 4 for x.
$2(4) + 3 \stackrel{?}{=} 11$
$8 + 3 \stackrel{?}{=} 11$
$11 = 11$ ✓

The variable is divided by 2 and then 3 is subtracted.
$\frac{x}{2} - 3 = 11$ To undo subtraction,
$+3 \quad +3$ add 3.
$\frac{x}{2} = 14$ To undo division,
$2 \cdot \frac{x}{2} = 2 \cdot 14$ multiply by 2.
$x = 28$
Check: Substitute 28 for x.
$\frac{28}{2} - 3 \stackrel{?}{=} 11$
$14 - 3 \stackrel{?}{=} 11$
$11 = 11$ ✓

Complete to solve and check each equation.

6. $3t + 7 = 19$ To undo addition,
   $-7 \quad -7$ subtract.
   $3t = 12$ To undo multiplication,
   $3t \div 3 = 12 \div 3$ divide.
   $t = 4$
   Check: $3t + 7 = 19$
   $3(\underline{4}) + 7 \stackrel{?}{=} 19$ Substitute for t.
   $\underline{12} + 7 \stackrel{?}{=} 19$
   $19 = 19$ ✓

7. $\frac{w}{3} - 7 = 5$ To undo subtraction,
   $+7 \quad +7$ add.
   $\frac{w}{3} = 12$ To undo division,
   $3 \cdot \frac{w}{3} = 3 \cdot 12$ multiply.
   $w = 36$
   Check: $\frac{w}{3} - 7 = 5$
   $\frac{36}{3} - 7 \stackrel{?}{=} 5$ Substitute.
   $12 - 7 \stackrel{?}{=} 5$
   $5 = 5$ ✓

8. $\frac{z - 3}{2} = 8$ To undo division,
   $2 \cdot \frac{z - 3}{2} = 2 \cdot 8$ multiply.
   $z - 3 = 16$ To undo subtraction,
   $+3 \quad +3$ add.
   $z = 19$
   Check: $\frac{z - 3}{2} = 8$
   $\frac{19 - 3}{2} \stackrel{?}{=} 8$ Substitute.
   $\frac{16}{2} \stackrel{?}{=} 8$
   $8 = 8$ ✓

LESSON 2-8 Challenge
Work It Algebraically!

An equation may be used to solve a problem involving probability.

A bag contains marbles of four colors: red, white, blue, and yellow. There are 3 more blue marbles than red, and 48 marbles in all. How many blue marbles are there if, in one draw, the probability of getting a blue marble is $\frac{5}{12}$?

Let x = the number of blue marbles.
$P(\text{blue}) = \frac{\text{number of successes}}{\text{total number}}$
$\frac{5}{12} = \frac{x}{48}$
$\frac{5}{12} \cdot 48 = 48 \cdot \frac{x}{48}$ Multiply by 48.
$20 = x$

So, there are 20 blue marbles in the bag.

Write and solve an equation for each problem.

1. A box has four kinds of candies: lime, orange, cherry, and mint. There are 9 more lime candies than orange, and 36 candies in all. How many lime candies are there if, in one draw, the probability of getting a lime candy is $\frac{4}{9}$?

   Let x = number of lime candies.
   $\frac{x}{36} = \frac{4}{9}$
   $36 \cdot \frac{x}{36} = \frac{4}{9} \cdot 36$
   $x = 16$
   There are __16__ lime candies.

2. A carton has four kinds of cookies: lemon, mint, vanilla, and chocolate. There are 7 fewer mint cookies than lemon, and 64 cookies in all. How many mint cookies are there if, in one draw, the probability of getting a lemon cookie is $\frac{5}{8}$?

   Let x = number of lemon cookies.
   Then $x - 7$ = number of mint cookies.
   $\frac{x}{64} = \frac{5}{8}$
   $64 \cdot \frac{x}{64} = \frac{5}{8} \cdot 64$
   $x = 40$ ← lemon cookies
   There are __33__ mint cookies.

LESSON 2-8 Problem Solving
Solving Two-Step Equations

The chart below describes three different long distance calling plans. Jamie has budgeted $20 per month for long distance calls. Write the correct answer.

Plan	Monthly Access Fee	Charge per minute
A	$3.95	$0.08
B	$8.95	$0.06
C	$0	$0.10

1. How many minutes will Jamie be able to use per month with plan A? Round to the nearest minute.
   201 min

2. How many minutes will Jamie be able to use per month with plan B? Round to the nearest minute.
   184 min

3. How many minutes will Jamie be able to use per month with plan C? Round to the nearest minute.
   200 min

4. Which plan is the best deal for Jamie's budget?
   Plan A

5. Nolan has budgeted $50 per month for long distance. Which plan is the best deal for Nolan's budget?
   Plan B

The table describes four different car loans that Susana can get to finance her new car. The total column gives the amount she will end up paying for the car including the down payment and the payments with interest. Choose the letter for the best answer.

Loan	Down Payment	Number of Months	Total
A	$2000	60	$19,821.20
B	$1000	48	$19,390.72
C	$0	60	$20,197.20

6. How much will Susana pay each month with loan A?
   A $252.04   C $330.35
   (B) $297.02   D $353.68

7. How much will Susana pay each month with loan B?
   F $300.85   H $323.17
   G $306.50   **(J) $383.14**

8. How much will Susana pay each month with loan C?
   (A) $336.62   C $369.95
   B $352.28   D $420.78

9. Which loan will give Susana the smallest monthly payment?
   (F) Loan A   H Loan C
   G Loan B   J They are equal

LESSON 2-8 Reading Strategies
Analyze Information

Break a problem into parts and analyze the information.

> Jill has $8 in her pocket now. She had $20 when she left for the movies. How much money did she spend?

Answer the questions in Exercises 1–4 to solve this problem.

1. How much money did Jill start with?
   $20

2. How much money does Jill have left?
   $8

3. What is the difference between these two amounts?
   $20 − $8 = $12

4. How much money did Jill spend?
   $12

> Mark paid $45 at the music store for 3 CDs and a pack of batteries, before tax. The batteries cost $6. How much did Mark pay for each of the CDs?

Answer the questions in Exercises 5–9 to solve this problem.

5. How much did Mark spend at the music store?
   $45

6. How much did Mark spend on batteries?
   $6

7. What is the difference between these two amounts?
   $45 − $6 = $39

8. Since Mark paid $39 for CDs, divide $39 by 3.
   $39 ÷ 3 = $13

9. How much did Mark pay for each CD?
   $13

LESSON 2-8 Puzzles, Twisters & Teasers
Have a Ball!

Solve the equations. Match the letters of the variables to the answers to solve the riddle.

1. $7 + \frac{l}{5} = -4$ $l = $ **−55**

2. $46 - 3t = -23$ $t = $ **23**

3. $8 = 6 + \frac{a}{4}$ $a = $ **8**

4. $6h + 24 = 0$ $h = $ **−4**

5. $9 = -5b - 23$ $b = $ **−6.4**

6. $15 - 3e = -6$ $e = $ **7**

7. $15w - 4 = 41$ $w = $ **3**

8. $6s + 3 = -27$ $s = $ **−5**

9. $14o - 17 = 39$ $o = $ **4**

10. $\frac{n}{-3} - 2 = 8$ $n = $ **−30**

Where do penguins go to dance?

A   T   H   E   S   N   O   W   B   A   L   L
23 −4 7   −5 −30 4   3   −6.4 8   −55